The Gospel of Mary and other apocryphal Gospels

THE GOSPEL OF MARY, PETER, THOMAS, THE BIRTH OF MARY and

THE ACTS OF PONTIUS PILATE

Printed in Scotts Valley, CA - USA.

Anonymous.

The Gospel of Mary and other Apocryphal Gospels. The Gospel of Mary, Peter, Thomas, The Birth of Mary and The Acts of Pontius Pilate. / Anonymous – 1st ed.

1. History 2. Religion

TABLE OF CONTENTS

THE GOSPEL OF MARY 5

THE GOSPEL OF PETER 11

THE GOSPEL OF THOMAS 17

THE GOSPEL OF THE BIRTH OF MARY 37

THE ACTS OF PONTIUS PILATE 51

THE GOSPEL OF MARY

This gospel of Mary was probably written between the years 100 and 150. It is a fragmentary text found in 1896 in a codex, known as Papyrus Berolinensis 8502 (and also as the Berlin Gnostic Codex or the Akhmim Codex), which was purchased in Cairo by a German scholar. Its first pages were not included in the acquisition. There are two other shorter fragments in Greek, Papyrus Oxyrhynchus L 3525 and Papyrus Rylands 463, which have been unearthed a few years later in archaeological excavations at Oxyrhynchus in Northern Egypt. The manuscript we call Gospel of Mary is sometimes called Gospel of Mary Magdalene (or Magdala) because scholars do not agree about the Mary mentioned in the text, although most of the scholars believe it refers to Mary Magdalene. Although the Gospel of Mary is sometimes included in the Nag Hammadi library, but the Nag Hammadi manuscripts founded in 1945 did not include the Gospel of Mary.

This is a translation from the Papyrus Berolinensis 8502. The text begins with a question about the future of our planet. Chapters 1-3 are lost.

Chapter 4

"... Will the matter be completely destroyed or not?"

The Savior said, "All natures, all formed things, all creatures exist in and with one another and will again finish going into their own roots, because the nature of matter is dissolved into the roots of its nature alone. He who has ears to hear, let him hear."

Peter said to him, "You have been explaining all things to us, tell us another thing: What is the sin of the world?"

The Savior said, "There is no such thing as sin, but you make a sin when you act in accordance to fornication, which is called 'sin.' For this reason the Good came among you, pursuing the good essence of each nature, to restore it to its root." He went on to say, "This is why you get ill and die [...] because you love what deceives you. Anyone who thinks should consider this.

"Matter gave birth to a passion which has no image because it derives from what is contrary to nature. A disturbing confusion then occurred in the whole body. For this reason I said to you, have courage, and if you are unhappy, still take courage over against the various forms of image of nature. He who has ears to hear, let him hear."

When the Blessed One said this, he greeted all of them, saying "Peace be with you. Receive my peace for yourselves. Take heed so that nobody deceives you with the words, 'Look here!' or 'Look there!' for the Son of Man is within you. Follow him; those who seek him will find him. Go, therefore, and preach the Gospel of the Kingdom. Do not lay down any law beyond the ones I gave you, nor promulgate laws as a lawgiver, otherwise you will be dominated by it." After he said these things, he departed from them.

Chapter 5

They grieved and wept greatly, saying, "How will we go to the whole world and preach the Gospel of the Kingdom of the Son of Man? If even He was not spared, how will we be spared?"

Then Mary stood up and greeted all and said to her brethren, "Do not weep and be distressed nor let your hearts be irresolute, for his grace will be with you all and will defend you. Let us rather praise his greatness, for he prepared us and made us into true human beings."

When Mary said this, their hearts became lighter, and they began to discuss the words of the [Savior].

Peter said to Mary, "Sister, we know that the Savior loved you more than all other women[1]. Tell us the words of the Savior which you remember, the things you know; and we do not know, because we have not heard them."

Mary answered, "I will tell you what is hidden from you." And she began to say the following words to them. "I saw the Lord in a vision and I said to him, 'Lord, I saw you today in a vision.' He answered to me, 'Blessed are you, since you did not waver at the sight of me. For where the mind is, there is your treasure' I said to him, 'Lord, the person who sees the vision, sees it through the soul or through the spirit?' The Savior answered, 'It sees neither through the soul nor through the spirit, but the mind, which is between the two, sees the vision, and it...'"

Pages 11-14 are missing (end of chapter 5, chapters 6,7)

Chapter 8

"... And Desire said, 'I did not see you go down; but now I see you rising. Why do you lie, since you belong to me?' The soul answered, 'I saw you, but you did not see me or recognize me; You mistook the garment I wore for my true self. And you did not recognize me' After it had said this, it went joyfully and gladly away. Again it came to the third power, Ignorance. This power asked the soul: 'Where are you going? You are bound in wickedness, you are bound indeed. Do not judge'. And the soul said, 'Why do you judge me, if I did not go into judgment? I have been bound, though I have not bound. I was not understood, but I understand that everything will be dissolved, both things from the earth and the heaven.' After the soul had left the third

[1] cf. John 11:5, Luke 10:38-42

power behind, it rose upward, and saw the fourth power, which had seven forms. The first form is darkness, the second desire, the third is ignorance, the fourth the zeal for death, the fifth is the kingdom of the flesh, the sixth is the foolish wisdom of the flesh, the seventh is wrathful wisdom. These are the seven powers of wrath. They asked the soul, 'Where are you coming from, killer of men, or where are you going, conqueror of space?' The soul answered, 'What seized me is dead; what surrounded me is overcome; my desire has come to an end and ignorance is dead. In a world I was saved from a world, and in a "type," from a "type" that is above and from the chain of the impotence of knowledge, which is temporal. From now on I will reach rest in silence, for the time of the season of the Aeon.'"

Chapter 9

When Mary had said this, she was silent, since she reached the end of the conversation she had with the Savior. But Andrew said to the brethren, 'Say what you think concerning what she said, because I do not believe that the Savior said this, for certainly these teachings are strange."

Peter also opposed her in regard to these matters and asked them about the Savior. "Did he then speak secretly with a woman, in preference to us, and not openly? Are all we to turn back and listen to her? Did he prefer her to us?"

Then Mary wept and said to Peter, "My brother Peter, what do you think? Do you think that I thought this up myself in my heart or that I am lying concerning the Savior?"

Levi said to Peter, "Peter, you are always been a wrathful person. Now I see that you are contending against the woman like the adversaries. But if the Savior made her worthy, who are you to reject her? Surely the Savior knew her very well. For this reason he loved her more than us. And we should rather be ashamed and become Perfect Men, do as he commanded us, and proclaim the gospel, without laying down a commandment or law different than the one which the Savior spoke." When Levi had said this, they all began to go out in order to proclaim him and preach.

THE GOSPEL OF PETER

Translation by Sam Gibson

This fragmentary gospel was found in 1886 in a tomb of a monk at Akhmim, in Egypt. Both the beginning and the end of this text are missing. The Gospel of Peter is quoted by many writers who lived in the second and third centuries.

1 ...but among the Jews, no one washed his hands, neither did Herod nor any one of his judges. Since they were unwilling to wash, Pilate stood up. 2 Then Herod the king ordered the Lord to be taken away, saying to them "Do what I commanded you to do to him."

2 Joseph stood there, the friend of Pilate and the Lord, and knowing that they were about to crucify him, he went to Pilate and asked for the body of the Lord for burial. 2 And Pilate sent to Herod and asked for his body. 3 And Herod replied, "Brother Pilate, even if no one had asked for him, we would have buried him since the Sabbath is drawing near. For it is written in the Law, "The sun must not set upon one who has been executed."' And he turned him over to the people on the day before the Unleavened Bread, their feast.

3 They took out the Lord and kept pushing him along as they ran; and they would say, "Let's drag the son of God since we have him in our power." 2 And they threw a purple robe around him and made him sit upon the judgment seat and said, "Judge justly, King of Israel." 3 And one of them brought a crown a thorns and set it on the Lord's head. 4 And others standing around spat in his eyes, and others slapped his face, while others poked him with a rod. Some kept flogging him as they said, "Let us pay proper respect to the son of God."

4 And they brought two criminals and crucified the Lord between them. But he kept silence, as one feeling no pain. 2 And when they set the cross upright, they wrote thereon: "This is the King of Israel." 3 And they laid his garments before him, and divided them among themselves and gambled for them. 4 But one of those criminals reproached them, saying, "We suffer for the evils which we have done; but this man which hath become the Savior of men, what has he done to you?" 5 And they were angry with him, and commanded that his legs should not be broken, that so he might die in torment.

5 Now it was midday and darkness prevailed over all Judea. They were troubled and in an agony lest the sun should have set for he still lived. For it is written that, "The sun should not set upon him that hath been executed." 2 And one of them said, "Give him vinegar and gall to drink." And they mixed it and gave it to him to drink. 3 And they fulfilled all things and brought their sins upon their own heads. 4 Now many went about with lamps, supposing that it was night, and they laid down. 5 And the Lord cried out aloud saying, "My power, my power, you have forsaken me." When he had said this, he was taken up. 6 And in the same hour the veil of the temple of Jerusalem was rent in two.

6 And then they pulled the nails from the hands of the Lord and laid him on the ground. And the whole earth was shaken, and there came a great fear on all. 2 Then the sun came out, and it was found to be the ninth hour. 3 Now the Jews rejoiced, and gave his body unto Joseph to bury it, because he had beheld the good things which he did. 4 And Joseph took the Lord and washed him and wrapped him in linen and brought him unto his own tomb, which is called the "Joseph's Garden."

7 Then the Jews and the elders and the priests, when they perceived how great evil they had done themselves, began to lament and to say,

"Woe unto our sins! The judgment and the end of Jerusalem is near!" 2 But I began weeping with my friends, and out of fear we would have hid ourselves for we were sought after by them as criminals, and as thinking to set the temple on fire. 3And beside all these things we were fasting, and we sat mourning and weeping night and day until the Sabbath.

8 But the scribes and Pharisees and elders gathered together, for they had heard that all the people were murmuring and beating their breasts, saying, "If these very great signs have come to pass at his death, he must have been innocent!" 2 And the elders were afraid and came unto Pilate, begging him and saying, 3 "Give us soldiers that we may guard his tomb for three days, lest his disciples come and steal him away and the people suppose that he is risen from the dead, and do us harm." 4And Pilate gave them Petronius the centurion with soldiers to watch the tomb. And the elders and scribes came with them unto the tomb. 5 All who were there with the soldiers rolled a great stone to the entrance of the tomb 6 and plastered seven seals on it. Then they pitched a tent there and kept watch.

9 Early in the morning, as the Sabbath dawned, there came a large crowd from Jerusalem and the surrounding areas to see the sealed tomb. 2 But during the night before the Lord's day dawned, as the soldiers were keeping guard two by two in every watch, there came a great sound in the sky, 3 and they saw the heavens opened and two men descend shining with a great light, and they drew near to the tomb. 4 The stone which had been set on the door rolled away by itself and moved to one side, and the tomb was opened and both of the young men went in.

10 Now when these soldiers saw that, they woke up the centurion and the elders (for they also were there keeping watch). 2 While they were yet telling them the things which they had seen, they saw three men come out of the tomb, two of them sustaining the other one, and a cross following after them. 3 The heads of the two they saw had heads that reached up to heaven, but the head of him that was led by them went beyond heaven. 4 And they heard a voice out of the heavens saying, "Have you preached unto them that sleep?" 5 The answer that was heard from the cross was, "Yes!"

11 Those men took counsel with each other and thought to go and report these things to Pilate. 2 And while they were thinking the heavens were opened again and a man descended and entered the tomb. 3 When those who were with the centurion saw that, they hurried to go by night to Pilate and left the tomb that they were watching. They told all what they had seen and were in great despair saying, "He was certainly the son of God!" 4 Pilate answered them, saying, I do not have the blood of the son of God on my hands. This was all your doing." 5 Then all they came and besought and pleaded with him to order the centurion and the soldiers to tell nothing of what they had seen. 6 "For," they said, "it is better for us to be guilty of the greatest sin before God, than to fall into the hands of the Jews and to be stoned." 7 Pilate therefore ordered the centurion and the soldiers that they should say nothing.

12 Early on the Lord's day, Mary of Magdala, a disciple of the Lord, was afraid of the Jews, for they were inflamed with rage, so she had not performed at the tomb of the Lord the things that are customary for women to do for their loved ones that have died. 2 She took with her some women friends and came unto the tomb where he had been laid. 3 And they feared lest the Jews would see them, and said, "Even if we were not able to weep and lament him on the day that he was crucified, let us do so now at his tomb. 4 But who will roll the stone away for us that is set upon the door of the tomb, so that we may enter in and sit beside him and do what needs to be done?" 5 The stone was indeed great. "We fear that someone might see us. And if we cannot roll the stone away, let us cast down at the door these things which we bring as a memorial of him, and we will weep and beat our breasts until we arrive home."

13 And they went and found the tomb open. They drew near to it and looked in and saw a young man sitting in the middle of the tomb; He had a fair countenance and was clad in very bright raiment. He said unto them, 2 Why are you here? Who do you seek? You're not looking for the one that was crucified? He is risen and is gone. If you don't believe it, look in and see the place where he was laid down, for he is not there. For he has risen and is gone to the place that he had come from. 3 Then the women fled in fear.

14 Now it was the last day of Unleavened Bread, and many were returning to their homes since the feast was ending. 2 But we, the twelve disciples of the Lord, continued weeping and mourning, and each one grieving for what had happened, left for his own home. 3 But I, Simon Peter, and Andrew my brother, took our fishing nets and went to the sea. With us was Levi, the son of Alphaeus, whom the Lord...

THE GOSPEL OF THOMAS

Translation by Stephen Patterson and Marvin Meyer

This gospel was found in the Nag Hammadi excavations in 1947-1949, in a Coptic manuscript that dates from the 4th century. This gospel was mentioned in the 3rd century. It begins this way:

These are the secret sayings that the living Jesus spoke and Didymos Judas Thomas recorded.

1. And he said, "Whoever discovers the interpretation of these sayings will not taste death."

2. Jesus said, "Those who seek should not stop seeking until they find. When they find, they will be disturbed. When they are disturbed, they will marvel, and will reign over all. [And after they have reigned they will rest.]"

3. Jesus said, "If your leaders say to you, 'Look, the (Father's) kingdom is in the sky,' then the birds of the sky will precede you. If they say to you, 'It is in the sea,' then the fish will precede you. Rather, the (Father's) kingdom is within you and it is outside you.

When you know yourselves, then you will be known, and you will understand that you are children of the living Father. But if you do not know yourselves, then you live in poverty, and you are the poverty."

4. Jesus said, "The person old in days won't hesitate to ask a little child seven days old about the place of life, and that person will live.

For many of the first will be last, and will become a single one."

5. Jesus said, "Know what is in front of your face, and what is hidden from you will be disclosed to you.

For there is nothing hidden that will not be revealed. [And there is nothing buried that will not be raised.]"

6. His disciples asked him and said to him, "Do you want us to fast? How should we pray? Should we give to charity? What diet should we observe?"

Jesus said, "Don't lie, and don't do what you hate, because all things are disclosed before heaven. After all, there is nothing hidden that will not be revealed, and there is nothing covered up that will remain undisclosed."

7. Jesus said, "Lucky is the lion that the human will eat, so that the lion becomes human. And foul is the human that the lion will eat, and the lion still will become human."

8. And he said, "The person is like a wise fisherman who cast his net into the sea and drew it up from the sea full of little fish. Among them the wise fisherman discovered a fine large fish. He threw all the little fish back into the sea, and easily chose the large fish. Anyone here with two good ears had better listen!"

9. Jesus said, "Look, the sower went out, took a handful (of seeds), and scattered (them). Some fell on the road, and the birds came and gathered them. Others fell on rock, and they didn't take root in the soil and didn't produce heads of grain. Others fell on thorns, and they choked the seeds and worms ate them. And others fell on good soil, and it produced a good crop: it yielded sixty per measure and one hundred twenty per measure."

10. Jesus said, "I have cast fire upon the world, and look, I'm guarding it until it blazes."

11. Jesus said, "This heaven will pass away, and the one above it will pass away.

The dead are not alive, and the living will not die. During the days when you ate what is dead, you made it come alive. When you are in the light, what will you do? On the day when you were one, you became two. But when you become two, what will you do?"

12. The disciples said to Jesus, "We know that you are going to leave us. Who will be our leader?"

Jesus said to them, "No matter where you are you are to go to James the Just, for whose sake heaven and earth came into being."

13. Jesus said to his disciples, "Compare me to something and tell me what I am like."

Simon Peter said to him, "You are like a just messenger."

Matthew said to him, "You are like a wise philosopher."

Thomas said to him, "Teacher, my mouth is utterly unable to say what you are like."

Jesus said, "I am not your teacher. Because you have drunk, you have become intoxicated from the bubbling spring that I have tended."

And he took him, and withdrew, and spoke three sayings to him. When Thomas came back to his friends they asked him, "What did Jesus say to you?"

Thomas said to them, "If I tell you one of the sayings he spoke to me, you will pick up rocks and stone me, and fire will come from the rocks and devour you."

14. Jesus said to them, "If you fast, you will bring sin upon yourselves, and if you pray, you will be condemned, and if you give to charity, you will harm your spirits.

When you go into any region and walk about in the countryside, when people take you in, eat what they serve you and heal the sick among them.

After all, what goes into your mouth will not defile you; rather, it's what comes out of your mouth that will defile you."

15. Jesus said, "When you see one who was not born of woman, fall on your faces and worship. That one is your Father."

16. Jesus said, "Perhaps people think that I have come to cast peace upon the world. They do not know that I have come to cast conflicts upon the earth: fire, sword, war.

For there will be five in a house: there'll be three against two and two against three, father against son and son against father, and they will stand alone."

17. Jesus said, "I will give you what no eye has seen, what no ear has heard, what no hand has touched, what has not arisen in the human heart."

18. The disciples said to Jesus, "Tell us, how will our end come?"

Jesus said, "Have you found the beginning, then, that you are looking for the end? You see, the end will be where the beginning is.

Congratulations to the one who stands at the beginning: that one will know the end and will not taste death."

19. Jesus said, "Congratulations to the one who came into being before coming into being.

If you become my disciples and pay attention to my sayings, these stones will serve you.

For there are five trees in Paradise for you; they do not change, summer or winter, and their leaves do not fall. Whoever knows them will not taste death."

20. The disciples said to Jesus, "Tell us what Heaven's kingdom is like."

He said to them, "It's like a mustard seed, the smallest of all seeds, but when it falls on prepared soil, it produces a large plant and becomes a shelter for birds of the sky."

21. Mary said to Jesus, "What are your disciples like?"

He said, "They are like little children living in a field that is not theirs. When the owners of the field come, they will say, 'Give us back our field.' They take off their clothes in front of them in order to give it back to them, and they return their field to them.

For this reason I say, if the owners of a house know that a thief is coming, they will be on guard before the thief arrives and will not let the thief break into their house (their domain) and steal their possessions.

As for you, then, be on guard against the world. Prepare yourselves with great strength, so the robbers can't find a way to get to you, for the trouble you expect will come.

Let there be among you a person who understands.

When the crop ripened, he came quickly carrying a sickle and harvested it. Anyone here with two good ears had better listen!"

22. Jesus saw some babies nursing. He said to his disciples, "These nursing babies are like those who enter the (Father's) kingdom."

They said to him, "Then shall we enter the (Father's) kingdom as babies?"

Jesus said to them, "When you make the two into one, and when you make the inner like the outer and the outer like the inner, and the upper like the lower, and when you make male and female into a single one, so that the male will not be male nor the female be female, when you make eyes in place of an eye, a hand in place of a hand, a foot in place of a foot, an image in place of an image, then you will enter [the kingdom]."

23. Jesus said, "I shall choose you, one from a thousand and two from ten thousand, and they will stand as a single one."

24. His disciples said, "Show us the place where you are, for we must seek it."

He said to them, "Anyone here with two ears had better listen! There is light within a person of light, and it shines on the whole world. If it does not shine, it is dark."

25. Jesus said, "Love your friends like your own soul, protect them like the pupil of your eye."

26. Jesus said, "You see the sliver in your friend's eye, but you don't see the timber in your own eye. When you take the timber out of your own eye, then you will see well enough to remove the sliver from your friend's eye."

27. "If you do not fast from the world, you will not find the (Father's) kingdom. If you do not observe the Sabbath as a Sabbath you will not see the Father."

28. Jesus said, "I took my stand in the midst of the world, and in flesh I appeared to them. I found them all drunk, and I did not find any of them thirsty. My soul ached for the children of humanity, because they are blind in their hearts and do not see, for they came into the world empty, and they also seek to depart from the world empty.

But meanwhile they are drunk. When they shake off their wine, then they will change their ways."

29. Jesus said, "If the flesh came into being because of spirit, that is a marvel, but if spirit came into being because of the body, that is a marvel of marvels.

Yet I marvel at how this great wealth has come to dwell in this poverty."

30. Jesus said, "Where there are three deities, they are divine. Where there are two or one, I am with that one."

31. Jesus said, "No prophet is welcome on his home turf; doctors don't cure those who know them."

32. Jesus said, "A city built on a high hill and fortified cannot fall, nor can it be hidden."

33. Jesus said, "What you will hear in your ear, in the other ear proclaim from your rooftops.

After all, no one lights a lamp and puts it under a basket, nor does one put it in a hidden place. Rather, one puts it on a lampstand so that all who come and go will see its light."

34. Jesus said, "If a blind person leads a blind person, both of them will fall into a hole."

35. Jesus said, "One can't enter a strong person's house and take it by force without tying his hands. Then one can loot his house."

36. Jesus said, "Do not fret, from morning to evening and from evening to morning, [about your food--what you're going to eat, or about your clothing--] what you are going to wear. [You're much better than the lilies, which neither card nor spin.

As for you, when you have no garment, what will you put on? Who might add to your stature? That very one will give you your garment.]"

37. His disciples said, "When will you appear to us, and when will we see you?"

Jesus said, "When you strip without being ashamed, and you take your clothes and put them under your feet like little children and trample then, then [you] will see the son of the living one and you will not be afraid."

38. Jesus said, "Often you have desired to hear these sayings that I am speaking to you, and you have no one else from whom to hear them. There will be days when you will seek me and you will not find me."

39. Jesus said, "The Pharisees and the scholars have taken the keys of knowledge and have hidden them. They have not entered nor have they allowed those who want to enter to do so.

As for you, be as sly as snakes and as simple as doves."

40. Jesus said, "A grapevine has been planted apart from the Father. Since it is not strong, it will be pulled up by its root and will perish."

41. Jesus said, "Whoever has something in hand will be given more, and whoever has nothing will be deprived of even the little they have."

42. Jesus said, "Be passersby."

43. His disciples said to him, "Who are you to say these things to us?"

"You don't understand who I am from what I say to you.

Rather, you have become like the Judeans, for they love the tree but hate its fruit, or they love the fruit but hate the tree."

44. Jesus said, "Whoever blasphemes against the Father will be forgiven, and whoever blasphemes against the son will be forgiven, but whoever blasphemes against the holy spirit will not be forgiven, either on earth or in heaven."

45. Jesus said, "Grapes are not harvested from thorn trees, nor are figs gathered from thistles, for they yield no fruit.

Good persons produce good from what they've stored up; bad persons produce evil from the wickedness they've stored up in their hearts, and say evil things. For from the overflow of the heart they produce evil."

46. Jesus said, "From Adam to John the Baptist, among those born of women, no one is so much greater than John the Baptist that his eyes should not be averted.

But I have said that whoever among you becomes a child will recognize the (Father's) kingdom and will become greater than John."

47. Jesus said, "A person cannot mount two horses or bend two bows.

And a slave cannot serve two masters, otherwise that slave will honor the one and offend the other.

Nobody drinks aged wine and immediately wants to drink young wine. Young wine is not poured into old wineskins, or they might break, and aged wine is not poured into a new wineskin, or it might spoil.

An old patch is not sewn onto a new garment, since it would create a tear."

48. Jesus said, "If two make peace with each other in a single house, they will say to the mountain, 'Move from here!' and it will move."

49. Jesus said, "Congratulations to those who are alone and chosen, for you will find the kingdom. For you have come from it, and you will return there again."

50. Jesus said, "If they say to you, 'Where have you come from?' say to them, 'We have come from the light, from the place where the light came into being by itself, established [itself], and appeared in their image.'

If they say to you, 'Is it you?' say, 'We are its children, and we are the chosen of the living Father.'

If they ask you, 'What is the evidence of your Father in you?' say to them, 'It is motion and rest.'"

51. His disciples said to him, "When will the rest for the dead take place, and when will the new world come?"

He said to them, "What you are looking forward to has come, but you don't know it."

52. His disciples said to him, "Twenty-four prophets have spoken in Israel, and they all spoke of you."

He said to them, "You have disregarded the living one who is in your presence, and have spoken of the dead."

53. His disciples said to him, "Is circumcision useful or not?"

He said to them, "If it were useful, their father would produce children already circumcised from their mother. Rather, the true circumcision in spirit has become profitable in every respect."

54. Jesus said, "Congratulations to the poor, for to you belongs Heaven's kingdom."

55. Jesus said, "Whoever does not hate father and mother cannot be my disciple, and whoever does not hate brothers and sisters, and carry the cross as I do, will not be worthy of me."

56. Jesus said, "Whoever has come to know the world has discovered a carcass, and whoever has discovered a carcass, of that person the world is not worthy."

57 Jesus said, "The Father's kingdom is like a person who has [good] seed. His enemy came during the night and sowed weeds among the good seed. The person did not let the workers pull up the weeds, but said to them, 'No, otherwise you might go to pull up the weeds and pull up the wheat along with them.' For on the day of the harvest the weeds will be conspicuous, and will be pulled up and burned."

58. Jesus said, "Congratulations to the person who has toiled and has found life."

59. Jesus said, "Look to the living one as long as you live, otherwise you might die and then try to see the living one, and you will be unable to see."

60. He saw a Samaritan carrying a lamb and going to Judea. He said to his disciples, "that person ... around the lamb." They said to him, "So that he may kill it and eat it." He said to them, "He will not eat it while it is alive, but only after he has killed it and it has become a carcass."

They said, "Otherwise he can't do it."

He said to them, "So also with you, seek for yourselves a place for rest, or you might become a carcass and be eaten."

61. Jesus said, "Two will recline on a couch; one will die, one will live."

Salome said, "Who are you mister? You have climbed onto my couch and eaten from my table as if you are from someone."

Jesus said to her, "I am the one who comes from what is whole. I was granted from the things of my Father."

"I am your disciple."

"For this reason I say, if one is whole, one will be filled with light, but if one is divided, one will be filled with darkness."

62. Jesus said, "I disclose my mysteries to those [who are worthy] of [my] mysteries.

63 Jesus said, "There was a rich person who had a great deal of money. He said, 'I shall invest my money so that I may sow, reap, plant, and fill my storehouses with produce, that I may lack nothing.' These were the things he was thinking in his heart, but that very night he died. Anyone here with two ears had better listen!"

64. Jesus said, "A person was receiving guests. When he had prepared the dinner, he sent his slave to invite the guests.

The slave went to the first and said to that one, 'My master invites you.' That one said, 'Some merchants owe me money; they are coming to me tonight. I have to go and give them instructions. Please excuse me from dinner.'

The slave went to another and said to that one, 'My master has invited you.' That one said to the slave, 'I have bought a house, and I have been called away for a day. I shall have no time.'

The slave went to another and said to that one, 'My master invites you.' That one said to the slave, 'My friend is to be married, and I am to arrange the banquet. I shall not be able to come. Please excuse me from dinner.'

The slave went to another and said to that one, 'My master invites you.' That one said to the slave, 'I have bought an estate, and I am going to collect the rent. I shall not be able to come. Please excuse me.'

The slave returned and said to his master, 'Those whom you invited to dinner have asked to be excused.' The master said to his slave, 'Go out on the streets and bring back whomever you find to have dinner.'

Buyers and merchants [will] not enter the places of my Father."

65. He said, "A [...] person owned a vineyard and rented it to some farmers, so they could work it and he could collect its crop from them. He sent his slave so the farmers would give him the vineyard's crop. They grabbed him, beat him, and almost killed him, and the slave returned and told his master. His master said, 'Perhaps he didn't know them.' He sent another slave, and the farmers beat that one as well. Then the master sent his son and said, 'Perhaps they'll show my son some respect.' Because the farmers knew that he was the heir to the vineyard, they grabbed him and killed him. Anyone here with two ears had better listen!"

66. Jesus said, "Show me the stone that the builders rejected: that is the keystone."

67. Jesus said, "Those who know all, but are lacking in themselves, are utterly lacking."

68. Jesus said, "Congratulations to you when you are hated and persecuted; and no place will be found, wherever you have been persecuted."

69. Jesus said, "Congratulations to those who have been persecuted in their hearts: they are the ones who have truly come to know the Father.

Congratulations to those who go hungry, so the stomach of the one in want may be filled."

70. Jesus said, "If you bring forth what is within you, what you have will save you. If you do not have that within you, what you do not have within you [will] kill you."

71. Jesus said, "I will destroy [this] house, and no one will be able to build it [...]."

72. A [person said] to him, "Tell my brothers to divide my father's possessions with me."

He said to the person, "Mister, who made me a divider?"

He turned to his disciples and said to them, "I'm not a divider, am I?"

73. Jesus said, "The crop is huge but the workers are few, so beg the harvest boss to dispatch workers to the fields."

74. He said, "Lord, there are many around the drinking trough, but there is nothing in the well."

75. Jesus said, "There are many standing at the door, but those who are alone will enter the bridal suite."

76. Jesus said, "The Father's kingdom is like a merchant who had a supply of merchandise and found a pearl. That merchant was prudent; he sold the merchandise and bought the single pearl for himself.

So also with you, seek his treasure that is unfailing, that is enduring, where no moth comes to eat and no worm destroys."

77. Jesus said, "I am the light that is over all things. I am all: from me all came forth, and to me all attained.

Split a piece of wood; I am there.

Lift up the stone, and you will find me there."

78. Jesus said, "Why have you come out to the countryside? To see a reed shaken by the wind? And to see a person dressed in soft clothes, [like your] rulers and your powerful ones? They are dressed in soft clothes, and they cannot understand truth."

79. A woman in the crowd said to him, "Lucky are the womb that bore you and the breasts that fed you."

He said to [her], "Lucky are those who have heard the word of the Father and have truly kept it. For there will be days when you will say, 'Lucky are the womb that has not conceived and the breasts that have not given milk.'"

80. Jesus said, "Whoever has come to know the world has discovered the body, and whoever has discovered the body, of that one the world is not worthy."

81. Jesus said, "Let one who has become wealthy reign, and let one who has power renounce it."

82. Jesus said, "Whoever is near me is near the fire, and whoever is far from me is far from the (Father's) kingdom."

83. Jesus said, "Images are visible to people, but the light within them is hidden in the image of the Father's light. He will be disclosed, but his image is hidden by his light."

84. Jesus said, "When you see your likeness, you are happy. But when you see your images that came into being before you and that neither die nor become visible, how much you will have to bear!"

85. Jesus said, "Adam came from great power and great wealth, but he was not worthy of you. For had he been worthy, [he would] not [have tasted] death."

86. Jesus said, "[Foxes have] their dens and birds have their nests, but human beings have no place to lay down and rest."

87. Jesus said, "How miserable is the body that depends on a body, and how miserable is the soul that depends on these two."

88. Jesus said, "The messengers and the prophets will come to you and give you what belongs to you. You, in turn, give them what you have, and say to yourselves, 'When will they come and take what belongs to them?'"

89. Jesus said, "Why do you wash the outside of the cup? Don't you understand that the one who made the inside is also the one who made the outside?"

90. Jesus said, "Come to me, for my yoke is comfortable and my lordship is gentle, and you will find rest for yourselves."

91. They said to him, "Tell us who you are so that we may believe in you."

He said to them, "You examine the face of heaven and earth, but you have not come to know the one who is in your presence, and you do not know how to examine the present moment."

92. Jesus said, "Seek and you will find.

In the past, however, I did not tell you the things about which you asked me then. Now I am willing to tell them, but you are not seeking them."

93. "Don't give what is holy to dogs, for they might throw them upon the manure pile. Don't throw pearls [to] pigs, or they might ... it [...]."

94. Jesus [said], "One who seeks will find, and for [one who knocks] it will be opened."

95. [Jesus said], "If you have money, don't lend it at interest. Rather, give [it] to someone from whom you won't get it back."

96. Jesus [said], "The Father's kingdom is like [a] woman. She took a little leaven, [hid] it in dough, and made it into large loaves of bread. Anyone here with two ears had better listen!"

97. Jesus said, "The [Father's] kingdom is like a woman who was carrying a [jar] full of meal. While she was walking along [a] distant road, the handle of the jar broke and the meal spilled behind her [along] the road. She didn't know it; she hadn't noticed a problem. When she reached her house, she put the jar down and discovered that it was empty."

98. Jesus said, "The Father's kingdom is like a person who wanted to kill someone powerful. While still at home he drew his sword and

thrust it into the wall to find out whether his hand would go in. Then he killed the powerful one."

99. The disciples said to him, "Your brothers and your mother are standing outside."

He said to them, "Those here who do what my Father wants are my brothers and my mother. They are the ones who will enter my Father's kingdom."

100. They showed Jesus a gold coin and said to him, "The Roman emperor's people demand taxes from us."

He said to them, "Give the emperor what belongs to the emperor, give God what belongs to God, and give me what is mine."

101. "Whoever does not hate [father] and mother as I do cannot be my [disciple], and whoever does [not] love [father and] mother as I do cannot be my [disciple]. For my mother [...], but my true [mother] gave me life."

102. Jesus said, "Damn the Pharisees! They are like a dog sleeping in the cattle manger: the dog neither eats nor [lets] the cattle eat."

103. Jesus said, "Congratulations to those who know where the rebels are going to attack. [They] can get going, collect their imperial resources, and be prepared before the rebels arrive."

104. They said to Jesus, "Come, let us pray today, and let us fast."

Jesus said, "What sin have I committed, or how have I been undone? Rather, when the groom leaves the bridal suite, then let people fast and pray."

105. Jesus said, "Whoever knows the father and the mother will be called the child of a whore."

106. Jesus said, "When you make the two into one, you will become children of Adam, and when you say, 'Mountain, move from here!' it will move."

107. Jesus said, "The (Father's) kingdom is like a shepherd who had a hundred sheep. One of them, the largest, went astray. He left the ninety-nine and looked for the one until he found it. After he had toiled, he said to the sheep, 'I love you more than the ninety-nine.'"

108. Jesus said, "Whoever drinks from my mouth will become like me; I myself shall become that person, and the hidden things will be revealed to him."

109. Jesus said, "The (Father's) kingdom is like a person who had a treasure hidden in his field but did not know it. And [when] he died he left it to his [son]. The son [did] not know about it either. He took over the field and sold it. The buyer went plowing, [discovered] the treasure, and began to lend money at interest to whomever he wished."

110. Jesus said, "Let one who has found the world, and has become wealthy, renounce the world."

111. Jesus said, "The heavens and the earth will roll up in your presence, and whoever is living from the living one will not see death."

Does not Jesus say, "Those who have found themselves, of them the world is not worthy"?

112. Jesus said, "Damn the flesh that depends on the soul. Damn the soul that depends on the flesh."

113. His disciples said to him, "When will the kingdom come?"

"It will not come by watching for it. It will not be said, 'Look, here!' or 'Look, there!' Rather, the Father's kingdom is spread out upon the earth, and people don't see it."

[Saying probably added to the original collection at a later date:] 114. Simon Peter said to them, "Make Mary leave us, for females don't deserve life."

Jesus said, "Look, I will guide her to make her male, so that she too may become a living spirit resembling you males. For every female who makes herself male will enter the kingdom of Heaven."

THE GOSPEL OF THE BIRTH OF MARY

CHAPTER 1

1 The Parentage of Mary. 7 Joachim her father, and Anna her mother, go to Jerusalem to the feast of the dedication. 9 Issachar, the high priest, reproaches Joachim for being childless.

THE blessed and ever glorious Virgin Mary, sprung from the royal race and family of David, was born in the city of Nazareth, and educated at Jerusalem, in the temple of the Lord.

2 Her father's name was Joachim, and her mother's Anna. The family of her father was of Galilee and the city of Nazareth. The family of her mother was of Bethlehem.

3 Their lives were plain and right in the sight of the Lord, pious and faultless before men; for they divided all their substance into three parts;

4 One of which they devoted to the temple and officers of the temple; another they distributed among strangers, and persons in poor circumstances; and the third they reserved for themselves and the uses of their own family.

5 In this manner they lived for about twenty years chastely, in the favor of God, and the esteem of men, without any children.

6 But they vowed, if God should favor them with any issue, they would

devote it to the service of the Lord; on which account they went at every feast in the year to the temple of the Lord.

7 And it came to pass, that when the feast of the dedication drew near, Joachim, with some others of his tribe, went up to Jerusalem, and at that time, Isachar was high-priest;

8 Who, when he saw Joachim along with the rest of his neighbors, bringing his offerings, despised both him and his offerings, and asked him,

9 Why he, who had no children, would presume to appear among those who had? Adding, that his offerings could never be acceptable to God, who was judged by him unworthy to have children; the Scripture having said, Cursed is every one who shall not beget a male in Israel.

10 He further said, that he ought first to be free from that curse by begetting some issue, and then come with his offerings into the presence of God.

11 But Joachim being much confounded with the shame of such reproach, retired to the shepherds who were with the cattle in their pastures;

12 For he was not inclined to return home, lest his neighbors, who were present and heard all this from the high-priest, should publicly reproach him in the same manner.

CHAPTER 2

1 An angel appears to Joachim, 9 and informs him that Anna shall conceive and bring forth a daughter, who shall be called Mary, 11 be

brought up in the temple, 12 and while yet a virgin, in a way unparalleled, bring forth the Son of God: 13 Gives him a sign, 14 and departs.

BUT when he had been there for some time, on a certain day when he was alone, the angel of the Lord stood by him with a prodigious light.

2 To whom, being troubled at the appearance, the angel who had appeared to him, endeavoring to compose him, said:

3 Be not afraid, Joachim, nor troubled at the sight of me, for I am an angel of the Lord sent by him to you, that I might inform you that your prayers are heard, and your alms ascended in the sight of God.

4 For he hath surely seen your shame, and heard you unjustly reproached for not having children: for God is the avenger of sin, and not of nature;

5 And so when he shuts the womb of any person, he does it for this reason, that he may in a more wonderful manner again open it, and that which is born appear to be not the product of lust, but the gift of God.

6 For the first mother of your nation, Sarah, was she not barren even till her eightieth year: and yet even in the end of her old age brought forth Isaac, in whom the promise was made of a blessing to all nations.

7 Rachel, also, so much in favor with God, and beloved so much by holy Jacob, continued barren for a long time, yet afterwards was the mother of Joseph, who was not only governor of Egypt, but delivered many nations from perishing with hunger.

8 Who among the judges was more valiant than Sampson, or more holy than Samuel? And yet both their mothers were barren.

9 But if reason will not convince you of the truth of my words, that there are frequent conceptions in advanced years, and that those who were barren have brought forth to their great surprise; therefore Anna your wife shall bring you a daughter, and you shall call her name Mary;

10 She shall, according to your vow, be devoted to the Lord from her infancy, and be filled with the Holy Ghost from her mother's womb;

11 She shall neither eat nor drink any thing which is unclean, nor shall her conversation be without among the common people, but in the temple of the Lord; that so she may not fall under any slander or suspicion of what is bad.

12 So in the process of her years, as she shall be in a miraculous manner born of one that was barren, so she shall, while yet a virgin, in a way unparalleled, bring forth the Son of the most High God, who shall, be called Jesus, and, according to the signification of his name, be the Savior of all nations.

13 And this shall be a sign to you of the things which I declare, namely, when you come to the golden gate of Jerusalem, you shall there meet your wife Anna, who being very much troubled that you returned no sooner, shall then rejoice to see you.

14 When the angel had said this, he departed from him.

CHAPTER 3

1 The angel appears to Anna; 2 tells her a daughter shall be born unto her, 3 devoted to the service of the Lord in the temple, 5, who, being a virgin, and not knowing man, shall bring forth the Lord, 6 and gives her a sign therefore. 8 Joachim and Anna meet, and rejoice, 10 and praise

the Lord. 11 Anna conceives, and brings forth a daughter called Mary.

AFTERWARDS the angel appeared to Anna his wife, saying; Fear not, neither think that which you see is a spirit;

2 For I am that angel who hath offered up your prayers and alms before God, and am now sent to you, that I may inform you, that a daughter will be born unto you, who shall be called Mary, and shall be blessed above all women.

3 She shall be, immediately upon her birth, full of the grace of the Lord, and shall continue during the three years of her weaning in her father's house, and afterwards, being devoted to the service of the Lord, shall not depart from the temple, till she arrive to years of discretion.

4 In a word, she shall there serve the Lord night and day in fasting and prayer, shall abstain from every unclean thing, and never know any man;

5 But, being an unparalleled instance without any pollution or defilement, and a virgin not knowing any man, shall ring forth a son, and a maid shall bring forth the Lord, who both by his grace and name and works, shall be the Savior of the world.

6 Arise therefore, and go up to Jerusalem, and when you shall come to that which is called the golden gate (because it is gilt with gold), as a sign of what I have told you, you shall meet your husband, for whose safety you have been so much concerned.

7 When therefore you find these things thus accomplished, believe that all the rest which I have told you, shall also undoubtedly be accomplished.

8 According therefore to the command of the angel, both of them left

the places where they were, and when they came to the place specified in the angels prediction, they met each other.

9 Then, rejoicing at each other's vision, and being fully satisfied in the promise of a child, they gave due thanks to the Lord, who exalts the humble.

10 After having praised the Lord, they returned home, and lived in a cheerful and assured expectation of the promise of God.

11 So Anna conceived, and brought forth a daughter, and, according to the angel's command, the parents did call her name Mary.

CHAPTER 4

1 Mary brought to the temple at three years old. 6 Ascends the stairs of the temple by miracle. 8 Her parents sacrifice and return home.

AND when three years were expired, and the time of her weaning complete, they brought the Virgin to the temple of the Lord with offerings.

2 And there were about the temple, according to the fifteen Psalms of degrees, fifteen stairs to ascend.

3 For the temple being built in a mountain, the altar of burnt- offering, which was without, could not be come near but by stairs;

4 The parents of the blessed Virgin and infant Mary put her upon one of these stairs;

5 But while they were putting off their clothes, in which they had

traveled, and according to custom putting on some that were more neat and clean,

6 In the mean time the Virgin of the Lord in such a manner went up all the stairs one after another, without the help of any to lead her or lift her, that any one would have judged from hence, that she was of perfect age.

7 Thus the Lord did, in the infancy of his Virgin, work this extraordinary work, and evidence by this miracle how great she was like to be hereafter.

8 But the parents having offered up their sacrifice, according to the custom of the law, and perfected their vow, left the Virgin with other virgins in the apartments of the temple, who were to be brought up there, and they returned home.

CHAPTER 5

2 Mary ministered unto by angels. 4 The high priest orders all virgins of fourteen years old to quit the temple and endeavor to be married. 5 Mary refuses, 6 having vowed her virginity to the Lord. 7 The high-priest commands a meeting of the chief persons of Jerusalem, 11 who seek the Lord for counsel in the matter. 13 A voice from the mercy-seat. 15 The high-priest obeys it by ordering all the unmarried men of the house of David to bring their rods to the altar, 17 that his rod which should flower, and on which the Spirit of God should sit, should betroth the Virgin.

BUT the Virgin of the Lord, as she advanced in years, increased also in perfections, and according to the saying of the Psalmist, her father and mother forsook her, but the Lord took care of her.

2 For she every day had the conversation of angels, and every day received visitors from God, which preserved her from all sorts of evil, and caused her to abound with all good things;

3 So that when at length she arrived to her fourteenth year, as the wicked could not lay any thing to her charge worthy of reproof, so all good persons, who were acquainted with her, admired her life and conversation.

4 At that time the high-priest made a public order, That all the virgins who had public settlements in the temple, and were come to this age, should return home, and, as they were now of a proper maturity, should, according to the custom of their country, endeavor to be married.

5 To which command, though all the other virgins readily yielded obedience, Mary the Virgin of the Lord alone answered, that she could not comply with it,

6 Assigning these reasons, that both she and her parents had devoted her to the service of the Lord; and besides, that she had vowed virginity to the Lord, which vow she was resolved never to break through by lying with a man.

7 The high-priest being hereby brought into a difficulty,

8 Seeing he durst neither on the one hand dissolve the vow, and disobey the Scripture, which says, Vow and pay,

9 Nor on the other hand introduce a custom, to which the people were strangers, commanded,

10 That at the approaching feast all the principal persons both of Jerusalem and the neighboring places should meet together, that he

might have their advice, how he had best proceed in so difficult a case.

11 When they were accordingly met, they unanimously agreed to seek the Lord, and ask counsel from him on this matter.

12 And when they were all engaged in prayer, the high-priest according to the usual way, went to consult God.

13 And immediately there was a voice from the ark, and the mercy seat, which all present heard, that it must be enquired or sought out by a prophecy of Isaiah, to whom the Virgin should be given and be betrothed;

14 For Isaiah saith, there shall come forth a rod out of the stem of Jesse, and a flower shall spring out of its root,

15 And the Spirit of the Lord shall rest upon him, the Spirit of Wisdom and Understanding, the Spirit of Counsel and Might, the Spirit of Knowledge and Piety, and the Spirit of the fear of the Lord shall fill him.

16 Then, according to this prophecy, he appointed, that all the men of the house and family of David, who were marriageable, and not married, should bring their several rods to the altar,

17 And out of whatsoever person's rod after it was brought, a flower should bud forth, and on the top of it the Spirit of the Lord should sit in the appearance of a dove, he should be the man to whom the Virgin should be given and be betrothed.

CHAPTER 6

1 Joseph draws back his rod. 5 The dove pitches on it. He betroths Mary and returns to Bethlehem. 7 Mary returns to her parents' house at Galilee.

AMONG the rest there was a man named Joseph of the house and family of David, and a person very far advanced in years, who kept back his rod, when every one besides presented his.

2 So that when nothing appeared agreeable to the heavenly voice, the high-priest judged it proper to consult God again.

3 Who answered that he to whom the Virgin was to be betrothed was the only person of those who were brought together, who had not brought his rod.

4 Joseph therefore was betrayed.

5 For, when he did bring his rod, and a dove coming from Heaven pitched upon the top of it, every one plainly saw, that the Virgin was to be betrothed to him.

6 Accordingly, the usual ceremonies of betrothing being over, he returned to his own city of Bethlehem, to set his house in order, and make the needful provisions for the marriage.

7 But the Virgin of the Lord, Mary, with seven other virgins of the same age, who had been weaned at the same time, and who had been appointed to attend her by the priest, returned to her parents' house in Galilee.

CHAPTER 7

7 The salutation of the Virgin by Gabriel, who explains to her that she shall conceive, without lying with a man, while a Virgin, 19 by the Holy Ghost coming upon her without the heats of lust. 21 She submits.

NOW at this time of her first coming into Galilee, the angel Gabriel was sent to her from God, to declare to her the conception of our Savior, and the manner and way of her conceiving him.

2 Accordingly going into her, he filled the chamber where she was with a prodigious light, and in a most courteous manner saluting her, he said,

3 Hail, Mary! Virgin of the Lord most acceptable! O Virgin full of grace! The Lord is with you. You are blessed above all women, and you are blessed above all men, that have been hitherto born.

4 But the Virgin, who had before been well acquainted with the countenances of angels, and to whom such light from heaven was no uncommon thing,

5 Was neither terrified with the vision of the angel, nor astonished at the greatness of the light, but only troubled about the angel's words,

6 And began to consider what so extraordinary a salutation should mean, what it did portend, or what sort of end it would have.

7 To this thought the angel, divinely inspired, replies;

8 Fear not, Mary, as though I intended anything inconsistent with your chastity in this salutation:

9 For you have found favor with the Lord, because you made virginity

your choice.

10 Therefore while you are a Virgin, you shall conceive without sin, and bring forth a son.

11 He shall be great, because he shall reign from sea to sea, and from the rivers even to the ends of the earth?

12 And he shall be called the Son of the Highest; for he who is born in a mean state on earth, reigns in an exalted one in heaven.

13 And the Lord shall give him the throne of his father David, and he shall reign over the house of Jacob for ever, and of his kingdom there shall be no end.

14 For he is the King of Kings, and Lord of Lords, and his throne is forever and ever.

15 To this discourse of the angel the Virgin replied, not, as though she were unbelieving, but willing to know the manner of it.

16 She said, How can that be? For seeing, according to my vow, I have never known any man, how can I bear a child without the addition of a man's seed.

17 To this the angel replied and said, Think not, Mary, that you shall conceive in the ordinary way.

18 For, without lying with a man, while a Virgin, you shall conceive; while a Virgin, you shall bring forth; and while a Virgin shall give suck.

19 For the Holy Ghost shall come upon you, and the power of the Most High shall overshadow you, without any of the heats of lust.

20 So that which shall be born of you shall be only holy, because it only is conceived without sin, and being born, shall be called the Son of God.

21 Then Mary stretching forth her hands, and lifting her eyes to heaven, said, Behold the handmaid of the Lord! Let it be unto me according to thy word.

CHAPTER 8

1 Joseph returns to Galilee, to marry the Virgin he had betrothed; 4 perceives she is with child, 5 is uneasy, 7 purposes to put her away privily, 8 is told by the angel of the Lord it is not the work of man but the Holy Ghost; 12 Marries her, but keeps chaste, 13 removes with her to Bethlehem, 15 where she brings forth Christ.

JOSEPH therefore went from Judea to Galilee, with intention to marry the Virgin who was betrothed to him:

2 For it was now near three months since she was betrothed to him.

3 At length it plainly appeared she was with child, and it could not be hid from Joseph:

4 For going to the Virgin in a free manner, as one espoused, and talking familiarly with her, he perceived her to be with child,

5 And thereupon began to be uneasy and doubtful, not knowing what course it would be best to take;

6 For being a just man, he was not willing to expose her, nor defame her by the suspicion of being a harlot, since he was a pious man:

7 He purposed therefore privately to put an end to their agreement, and as privately to send her away.

8 But while he was meditating these things, behold the angel of the Lord appeared to him in his sleep, and said, Joseph, son of David, fear not;

9 Be not willing to entertain any suspicion of the Virgin's being guilty of fornication, or to think any thing amiss of her, neither be afraid to take her to wife:

10 For that which is begotten in her and now distresses your mind, is not the work of man, but the Holy Ghost.

11 For she of all women is that only Virgin who shall bring forth the Son of God, and you shall call his name Jesus, that is, Savior: for he will save his people from their sins.

12 Joseph thereupon, according to the command of the angel, married the Virgin, and did not know her, but kept her in chastity.

13 And now the ninth month from her conception drew near, when Joseph took his wife and what other things were necessary to Bethlehem, the city from whence he came.

14 And it came to pass, while they were there, the days were fulfilled for her bringing forth.

15 And she brought forth her first-born son, as the holy Evangelists have taught, even our Lord Jesus Christ, who with the Father, Son, and Holy Ghost, lives and reigns to everlasting ages.

THE ACTS OF PONTIUS PILATE

also called

THE GOSPEL OF NICODEMUS

The Gospel concerning the Sufferings and Resurrection of our Master and Savior, JESUS CHRIST.

CHAPTER I.

1 Christ accused to Pilate by the Jews of healing on the Sabbath. 9 Summoned before Pilate by a messenger who does him honor. 20 Worshipped by the standards bowing down to him.

ANNAS and Caiphas, and Summas, and Datam, Gamaliel, Judas, Levi, Nepthalim, Alexander, Cyrus, and other Jews, went to Pilate about Jesus, accusing him with many bad crimes.

2 And said, We are assured that Jesus is the son of Joseph, the carpenter, and born of Mary, and that he declares himself the Son of God, and a king; and not only so, but attempts the dissolution of the Sabbath, and the laws of our fathers.

3 Pilate replied, What is it which he declares? and what is it which he attempts dissolving?

4 The Jews told him, We have a law which forbids doing cures on the

Sabbath day; but he cures both the lame and the deaf, those afflicted with the palsy, the blind, the lepers, and demoniacs, on that day, by wicked methods.

5 Pilate replied, How can he do this by wicked methods? They answered He is a conjurer, and casts out devils by the prince of the devils; and so all things, become subject to him.

6 Then said Pilate, Casting out devils seems not to be the work of an unclean spirit, but to proceed from the power of God.

7 The Jews replied to Pilate, We entreat your highness to summon him to appear before your tribunal, and hear him yourself.

8 Then Pilate called a messenger, and said to him, By what means will Christ be brought hither?

9 Then went the messenger forth, and knowing Christ, worshipped him; and having spread the cloak which he had in his hand upon the ground, he said, Lord, walk upon this, and go in, for the governor calls thee.

10 When the Jews perceived what the messenger had done, they exclaimed (against him) to Pilate, and said, Why did you not give him his summons by a beadle, and not by a messenger?--For the messenger, when he saw him, worshipped him, and spread the cloak which he had in his hand upon the ground before him, and said to him, Lord, the governor calls thee.

11 Then Pilate called the messenger, and said, Why hast thou done thus?

12 The messenger replied, When thou sentest me from Jerusalem to Alexander, I saw Jesus sitting in a mean figure upon a she-ass, and the

children of the Hebrews cried out, Hosannah, holding boughs of trees in their hands.

13 Others spread their garments in the way, and said, Save us, thou who art in heaven; blessed is he who cometh in the name of the Lord.

14 Then the Jews cried out, against the messenger, and said, The children of the Hebrews made their acclamations in the Hebrew language; and how couldst thou, who art a Greek, understand the Hebrew?

15 The messenger answered them and said, I asked one of the Jews and said, What is this which the children do cry out in the Hebrew language?

16 And he explained it to me, saying, they cry out, Hosannah, which being interpreted, is, O Lord, save me; or, O Lord, save.

17 Pilate then said to them, Why do you yourselves testify to the words spoken by the children, namely, by your silence? In what has the messenger done amiss? And they were silent.

18 Then the governor said unto the messenger, Go forth and endeavor by any means to bring him in.

19 But the messenger went forth and did as before; and said, Lord come in, for the governor calleth thee.

20 And as Jesus was going in by the ensigns, who carried the standards, the tops of them bowed down and worshipped Jesus.

21 Whereupon the Jews exclaimed more vehemently against the ensigns.

22 But Pilate said to the Jews, I know it is not pleasing to you that the tops of the standards did of themselves bow and worship Jesus; but why do ye exclaim against the ensigns, as if they had bowed and worshipped?

23 They replied to Pilate, We saw the ensigns themselves bowing and worshipping Jesus.

24 Then the governor called the ensigns, and said unto them, Why did you do thus?

25 The ensigns said to Pilate, We are all Pagans and worship the gods in temples; and how should we think anything about worshipping him? We only held the standards in our hands, and they bowed themselves and worshipped him.

26 Then said Pilate to the rulers of the synagogue, Do ye yourselves choose some strong men, and let them hold the standards, and we shall see whether they will then bend of themselves.

27 So the elders of the Jews sought out twelve of the most strong and able old men, and made them hold the standards, and they stood in the presence of the governor.

28 Then Pilate said to the messenger, Take Jesus out, and by some means bring him in again. And Jesus and the messenger went out of the hall.

29 And Pilate called the ensigns who before had borne the standards, and swore to them, that if they had not borne the standards in that manner when Jesus before entered in, he would cut off their heads.

30 Then the governor commanded Jesus to come in again.

31 And the messenger did as he had done before, and very much entreated Jesus that he would go upon his cloak, and walk on it; and he did walk upon it, and went in.

32 And when Jesus went in, the standards bowed themselves as before, and worshipped him.

CHAPTER II.

2 Is compassionated by Pilate's wife, 7 charged with being born in fornication. 12 Testimony to the betrothing of his parents. 15 Hatred of the Jews to him.

NOW when Pilate saw this, he was afraid, and was about to rise from his seat.

2 But while he thought to rise, his own wife who stood at a distance, sent to him, saying, Have thou nothing to do with that just man; for I have suffered much concerning him in a vision this night.

3 When the Jews heard this they said to Pilate, Did we not say unto thee, He is a conjuror? Behold, he hath caused thy wife to dream.

4 Pilate then calling Jesus, said, thou hast heard what they testify against thee, and makest no answer?

5 Jesus replied, If they had not a power of speaking, they could not have spoke; but because every one has the command of his own tongue, to speak both good and bad, let him look to it.

6 But the elders of the Jews answered, and said to Jesus, What shall we look to?

7 In the first place, we know this concerning thee, that thou wast born

through fornication; secondly, that upon the account of thy birth the infants were slain in Bethlehem; thirdly, that thy father and mother Mary fled into Egypt, because they could not trust their own people.

8 Some of the Jews who stood by spake more favorably, We cannot say that he was born through fornication; but we know that his mother Mary was betrothed to Joseph, and so he was not born through fornication.

9 Then said Pilate to the Jews who affirmed him to be born through fornication, This your account is not true, seeing there was a betrothment, as they testify who are of your own nation.

10 Annas and Caiphas spake to Pilate, All this multitude of people is to be regarded, who cry out, that he was born through fornication, and is a conjurer; but they who deny him to be born through fornication, are his proselytes and disciples.

11 Pilate answered Annas and Caiphas, Who are the proselytes? They answered, They are those who are the children of Pagans, and are not become Jews, but followers of him.

12 Then replied Eleazer, and Asterius, and Antonius, and James, Caras and Samuel, Isaac and Phinees, Crispus and Agrippa, Annas and Judas, We are not proselytes, but children of Jews, and speak the truth, and were present when Mary was betrothed.

13 Then Pilate addressing himself to the twelve men who spake this, said to them, I conjure you by the life of Caesar, that ye faithfully declare whether he was born through fornication, and those things be true which ye have related.

14 They answered Pilate, We have a law whereby we are forbid to swear, it being a sin: Let them swear by the life of Caesar that it is not as

we have said, and we will be contented to be put to death.

15 Then said Annas and Caiphas to Pilate, Those twelve men will not believe that we know him to be basely born, and to be a conjurer, although he pretends that he is the Son of God, and a king: which we are so far from believing, that we tremble to hear.

16 Then Pilate commanded every one to go out except the twelve men who said he was not born through fornication, and Jesus to withdraw to a distance, and said to them, Why have the Jews a mind to kill Jesus?

17 They answered him, They are angry because he wrought cures on the sabbath day. Pilate said, Will they kill him for a good work? They say unto him, Yes, Sir.

CHAPTER III.

1 Is exonerated by Pilate. 11 Disputes with Pilate concerning truth.

THEN Pilate, filled with anger, went out of the hall, and said to the Jews, I call the whole world to witness that I find no fault in that man.

2 The Jews replied to Pilate, If he had not been a wicked person, we had not brought him before thee.

3 Pilate said to them, Do ye take him and try him by your law.

4 Then the Jews said, It is not lawful for us to put any one to death.

5 Pilate said to the Jews, The command, therefore, thou shalt not kill, belongs to you, but not to me.

6 And he went again into the hall, and called Jesus by himself, and said to him, Art thou the king of the Jews?

7 And Jesus answering, said to Pilate, Dost thou speak this of thyself, or did the Jews tell it thee concerning me?

8 Pilate answering, said to Jesus, Am I a Jew? The whole nation and rulers of the Jews have delivered thee up to me. What hast thou done?

9 Jesus answering, said, My kingdom is not of this world: if my kingdom were of this world, then would my servants fight, and I should not have been delivered to the Jews: but now my kingdom is not from hence.

10 Pilate said, Art thou a king then? Jesus answered, Thou sayest that I am a king: to this end was I born, and for this end came I into the world; and for this purpose I came, that I should bear witness to the truth; and every one who is of the truth, heareth my voice.

11 Pilate saith to him, What is truth?

12 Jesus said, Truth is from heaven.

13 Pilate said, Therefore truth is not on earth.

14 Jesus saith to Pilate, Believe that truth is on earth among those, who when they have the power of judgment, are governed by truth, and form right judgment.

CHAPTER IV.

1 Pilate finds no fault in Jesus. 16 The Jews demand his crucifixion.

THEN Pilate left Jesus in the hall, and went out to the Jews, and said, I find not any one fault in Jesus.

2 The Jews say unto him, But he said, I can destroy the temple of God,

and in three days build it up again.

3 Pilate saith to them, What sort of temple is that of which he speaketh?

4 The Jews say unto him, That which Solomon was forty-six years in building, he said he would destroy, and in three days build up.

5 Pilate said to them again, I am innocent from the blood of that man! do ye look to it.

6 The Jews say to him, His blood be upon us and our children. Then Pilate calling together the elders and scribes, priests and Levites, saith to them privately, Do not act thus; I have found nothing in your charge (against him) concerning his curing sick persons, and breaking the sabbath, worthy of death.

7 The priests and Levites replied to Pilate, By the life of Caesar, if any one be a blasphemer, he is worthy of death; but this man hath blasphemed against the Lord.

8 Then the governor again commanded the Jews to depart out of the hall; and calling Jesus, said to him, What shall I do with thee?

9 Jesus answered him, Do according as it is written.

10 Pilate said to him, How is it written?

11 Jesus saith to him, Moses and the prophets have prophesied concerning my suffering and resurrection.

12 The Jews hearing this, were provoked, and said to Pilate, Why wilt thou any longer hear the blasphemy of that man?

13 Pilate saith to them, If these words seem to you blasphemy, do ye take him, bring him to your court, and try him according to your law.

14 The Jews reply to Pilate, Our law saith, he shall be obliged to receive nine and thirty stripes, but if after this manner he shall blaspheme against the Lord, he shall be stoned.

15 Pilate saith unto them, If that speech of his was blasphemy, do ye try him according to your law.

16 The Jews say to Pilate, Our law command us not to put any one to death. We desire that he may be crucified, because he deserves the death of the cross.

17 Pilate saith to them, It is not fit he should be crucified: let him be only whipped and sent away.

18 But when the governor looked upon the people that were present and the Jews, he saw many of the Jews in tears, and said to the chief priests of the Jews, All the people do not desire his death.

19 The elders of the Jews answered to Pilate, We and all the people came hither for this very purpose, that he should die.

20 Pilate saith to them, Why should he die?

21 They said to him, Because he declares himself to be the Son of God and a King.

CHAP. V.

1 Nicodemus speaks in defense of Christ, and relates his miracles. 12 Another Jew, 26 with Veronica, 34 Centurio, and others, testify of

other miracles.

BUT Nicodemus, a certain Jew, stood before the governor, and said, I entreat thee, O righteous judge, that thou wouldst favor me with the liberty of speaking a few words.

2 Pilate said to him, Speak on.

3 Nicodemus said, I spake to the elders of the Jews, and the scribes, and priests and Levites, and all the multitude of the Jews, in their assembly; What is it ye would do with this man?

4 He is a man who hath wrought many useful and glorious miracles, such as no man on earth ever wrought before, nor will ever work. Let him go, and do him no harm; if he cometh from God, his miracles, (his miraculous cures) will continue; but if from men, they will come to nought.

5 Thus Moses, when he was sent by God into Egypt, wrought the miracles which God commanded him, before Pharaoh king of Egypt; and though the magicians of that country, Jannes and Jambres, wrought by their magic the same miracles which Moses did, yet they could not work all which he did;

6 And the miracles which the magicians wrought, were not of God, as ye know, O Scribes and Pharisees; but they who wrought them perished, and all who believed them.

7 And now let this man go; because the very miracles for which ye accuse him, are from God; and he is not worthy of death.

8 The Jews then said to Nicodemus, Art thou become his disciple, and making speeches in his favor?

9 Nicodemus said to them, Is the governor become his disciple also, and does he make speeches for him? Did not Caesar place him in that high post?

10 When the Jews heard this they trembled, and gnashed their teeth at Nicodemus, and said to him, Mayest thou receive his doctrine for truth, and have thy lot with Christ!

11 Nicodemus replied, Amen; I will receive his doctrine, and my lot with him, as ye have said.

12 Then another certain Jew rose up, and desired leave of the governor to hear him a few words.

13 And the governor said, Speak, what thou hast a mind.

14 And he said, I lay for thirty- eight years by the sheep-pool at Jerusalem, laboring under a great infirmity, and waiting for a cure which should be wrought by the coming of an angel, who at a certain time troubled the water: and whosoever first after the troubling of the water stepped in, was made whole of whatsoever disease he had.

15 And when Jesus saw me languishing there, he said to me, Wilt thou be made whole? And I answered, Sir, I have no man, when the water is troubled, to put me into the pool.

16 And he said unto me, Rise, take up thy bed and walk. And I was immediately made whole, and took up my bed and walked.

17 The Jews then said to Pilate, Our Lord Governor, pray ask him what day it was on which he was cured of his infirmity.

18 The infirm person replied, It was on the sabbath.

19 The Jews said to Pilate, Did we not say that he wrought his cures on the sabbath, and cast out devils by the prince of devils?

20 Then another certain Jew came forth, and said, I was blind, could hear sounds, but could not see any one; and as Jesus was going along, I heard the multitude passing by, and I asked what was there?

21 They told me that Jesus was passing by: then I cried out, saying, Jesus, Son of David, have mercy on me. And he stood still, and commanded that I should be brought to him, and said to me, What wilt thou?

22 I said, Lord, that I may receive my sight.

23 He said to me, Receive thy sight: and presently I saw, and followed him, rejoicing and giving thanks,

24 Another Jew also came forth, and said, I was a leper, and he cured me by his word only, saying, I will, be thou clean; and presently I was cleansed from my leprosy.

25 And another Jew came forth, and said I was crooked, and he made me straight by his word.

26 And a certain woman named Veronica, said, I was afflicted with an issue of blood twelve years, and I touched the hem of his garment, and presently the issue of blood stopped.

27 The Jews then said, We have a law, that a woman shall not be allowed as an evidence.

28 And, after other things, another Jew said, I saw Jesus invited to a wedding with his disciples, and there was a want of wine in Cana of Galilee;

29 And when the wine was all drank, he commanded the servants that they should fill six pots which were there with water, and they filled them up to the brim, and he blessed them and turned the water into wine, and all the people drank, being surprised at this miracle,

30 And another Jew stood forth, and said, I saw Jesus teaching in the synagogue at Capernaum; and there was in the synagogue a certain man who had a devil; and he cried out, saying, let me alone; what have we to do with thee, Jesus of Nazareth? Art thou come to destroy us? I know that thou art the Holy One of God.

31 And Jesus rebuked him, saying, Hold thy peace, unclean spirit, and come out of the man; and presently he came out of him, and did not at all hurt him.

32 The following things were also said by a Pharisee: I saw that a great company came to Jesus from Galilee and Judea, and the sea-cost, and many countries about Jordan; and many infirm persons came to him, and he healed them all.

33 And I heard the unclean spirits crying out, and saying, Thou art the Son of God. And Jesus strictly charged them, that they should not make him known.

34 After this another person, whose name was Centurio, said, I saw Jesus in Capernaum, and I entreated him, saying, Lord, my servant lieth at home sick of the palsy.

35 And Jesus said to me, I will come and cure him.

36 But I said, Lord, I am not worthy that thou shouldst come under my roof; but only speak the word, and my servant shall be healed.

37 And Jesus said unto me, Go thy way; and as thou hast believed so be

it done unto thee. And my servant was healed from that same hour.

38 Then a certain nobleman said, I had a son in Capernaum, who lay at the point of death; and when I heard that Jesus was come into Galilee, I went and besought him that he would come down to my house, and heal my son, for he was at the point of death.

39 He said to me, Go thy way, thy son liveth.

40 And my son was cured from that hour.

41 Besides these, also many others of the Jews, both men and Women, cried out and said, He is truly the Son of God, who cures all diseases only by his word, and to whom the devils are altogether subject.

42 Some of them farther said, This power can proceed from none but God.

43 Pilate said to the Jews, Why are not the devils subject to your doctors?

44 Some of them said, The power of subjecting devils cannot proceed but from God.

45 But others said to Pilate, That he had raised Lazarus from the dead, after he had been four days in his grave.

46 The governor hearing this, trembling, said to the multitude of the Jews, What will it profit you to shed innocent blood?

CHAPTER VI.

1 Pilate dismayed by the turbulence of the Jews, 5 who demand Barabbas to be released, and Christ to be crucified. 9 Pilate warmly

expostulates with them, 20 washes his hands of Christ's blood, 23 and sentences him to be whipped and crucified.

THEN Pilate having called together Nicodemus, and the fifteen men who said that Jesus was not born through fornication, said to them, What shall I do, seeing there is like to be a tumult among the people.

2 They say unto him, We know not; let them look to it who raise the tumult.

3 Pilate then called the multitude again, and said to them, Ye know that ye have a custom, that I should release to you one prisoner at the feast of the Passover:

4 I have a noted prisoner, a murderer, who is called Barabbas, and Jesus who is called Christ, in whom I find nothing that deserves death; which of them, therefore, have you a mind that I should release to you?

5 They all cry out, and say, Release to us Barabbas.

6 Pilate saith to them, What then shall I do with Jesus who is called Christ?

7 They all answer, Let him be crucified.

8 Again they cry out and say to Pilate, You are not the friend of Caesar, if you release this man; for he hath declared that he is the Son of God, and a king. But are you inclined that he should be king, and not Caesar?

9 Then Pilate filled with anger said to them, Your nation hath always been seditious, and you are always against those who have been serviceable to you.

10 The Jews replied, Who are those who have been serviceable to us?

11 Pilate answered them, Your God who delivered you from the hard bondage of the Egyptians, and brought you over the Red Sea as though it had been dry land, and fed you in the wilderness with manna and the flesh of quails, and brought water out of the rock, and gave you a law from heaven.

12 Ye provoked him all ways, and desired for yourselves a molten calf, and worshipped it, and sacrificed to it, and said, These are thy Gods, O Israel, which brought thee out of the land of Egypt:

13 On account of which your God was inclined to destroy you; but Moses interceded for you, and your God heard him, and forgave your iniquity.

14 Afterwards ye were enraged against, and would have killed your prophets, Moses and Aaron, when they fled to the tabernacle, and ye were always murmuring against God and his prophets.

15 And arising from his judgment seat, he would have gone out; but the Jews all cried out, We acknowledge Caesar to be king, and not Jesus;

16 Whereas this person, as soon as he was born, the wise men came and offered gifts unto him; which when Herod heard, he was exceedingly troubled, and would have killed him:

17 When his father knew this, he fled with him and his mother Mary into Egypt. Herod, when he heard he was born, would have slain him; and accordingly sent and slew all the children which were in Bethlehem, and in all the coasts thereof, from two years old and under.

18 When Pilate heard this account, he was afraid; and commanding

silence among the people, who made a noise, he said to Jesus, Art thou therefore a king?

19 All the Jews replied to Pilate, he is the very person whom Herod sought to have slain.

20 Then Pilate taking water, washed his hands before the people and said, I am innocent of the blood of this just person; look ye to it.

21 The Jews answered and said, His blood be upon us and our children.

22 Then Pilate commanded Jesus to be brought before him, and spake to him in the following words;

23 Thy own nation hath charged thee as making thyself a king; wherefore I, Pilate, sentence thee to be whipped according to the laws of former governors; and that thou be first bound, then hanged upon a cross in that place where thou art now a prisoner; and also two criminals with thee, whose names are Dimas and Gestas.

CHAP. VII.

1 Manner of Christ's crucifixion with the two thieves.

THEN Jesus went out of the hall, and the two thieves with him.

2 And when they came to the place which is called Golgotha, they stript him of his raiment, and girt him about with a linen cloth, and put a crown of thorns upon his head, and put a reed in his hand.

3 And in like manner did they to the two thieves who were crucified with him, Dimas on his right hand and Gestas on his left.

4 But Jesus said, My Father, forgive them, For they know not what they do.

5 And they divided his garments, and upon his vesture they cast lots.

6 The people in the mean time stood by, and the chief priests and elders of the Jews mocked him, saying, He saved others, let him now save himself if he can; if he be the son of God, let him now come down from the cross.

7 The soldiers also mocked him, and taking vinegar and gall, offered it to him to drink, and said to him, If thou art king of the Jews, deliver thyself.

8 Then Longinus, a certain soldier, taking a spear,' pierced his side, and presently there came forth blood and water.

9 And Pilate wrote the title upon the cross in Hebrew, Latin, and Greek letters, viz., THIS IS THE KING OF THE JEWS.

10 But one of the two thieves who were crucified with Jesus, whose name was Gestas, said to Jesus, If thou art the Christ, deliver thyself and us.

11 But the thief who was crucified on his right hand, whose name was Dimas, answering, rebuked him, and said, Dost not thou fear God, who art condemned to this punishment? We indeed receive rightly and justly the demerit of our actions; but this Jesus, what evil hath he done.

12 After this, groaning, he said to Jesus, Lord, remember me when thou comest into thy kingdom.

13 Jesus answering, said to him, Verily I say unto thee, that this day thou shalt be with me in Paradise.

CHAPTER VIII.

1 Miraculous appearance at his death. 10 The Jews say the eclipse was natural. 12 Joseph of Arimathcea embalms Christ's body and buries it.

AND it was about the sixth hour, and darkness was upon the face of the whole earth until the ninth hour.

2 And while the sun was eclipsed, behold the veil of the temple was rent from the top, to the bottom; and the rocks also were rent, and the graves opened, and many bodies of saints, which slept, arose.

3 And about the ninth hour Jesus cried out with a loud voice, Eli, Eli, lama sabacthani? which being interpreted is, My God, My God, why hast thou forsaken me?

4 And after these things, Jesus said, Father, into thy hands I commend my spirit; and having said this, he gave up the ghost.

5 But when the centurion saw that Jesus thus crying out gave up the ghost, he glorified God, and said, Of a truth this was a just man.

6 And all the people who stood by, were exceedingly troubled at the sight; and reflecting upon what had passed, smote upon their breasts, and then returned to the city of Jerusalem.

7 The centurion went to the governor, and related to him all that had passed:

8 And when he had heard all these things, he was exceedingly sorrowful;

9 And calling the Jews together, said to them, Have ye seen the miracle of the sun's eclipse, and the other things which came to pass, while

Jesus was dying?

10 Which when the Jews heard, they answered to the governor, The eclipse of the sun happened according to its usual custom.

11 But all those who were the acquaintance of Christ, stood at a distance, as did the women who had followed Jesus from Galilee, observing all these things.

12 And behold a certain man of Arimathaea, named Joseph, who was also a disciple of Jesus, but not openly so, for fear of the Jews, came to the governor, and entreated the governor that he would give him leave to take away the body of Jesus from the cross.

13 And the governor gave him leave.

14 And Nicodemus came, bringing with him a mixture of myrrh and aloes about a hundred pounds weight; and they took down Jesus from the cross with tears, and bound him in linen cloths with spices, according to the custom of burying among the Jews;

15 And placed him in a new tomb, which Joseph had built, and caused to be cut out of a rock, in which never any man had been put; and they rolled a great stone to the door of the sepulcher.

CHAPTER IX.

1 The Jews angry with Nicodemus: 5 and with, Joseph of Arimathaea, 7 whom they imprison.

WHEN the unjust Jews heard that Joseph had begged and buried the body of Jesus, they sought after Nicodemus, and those fifteen men who had testified before the governor, that Jesus was not born through fornication, and other good persons who had shown any good actions towards him.

2 But when they all concealed themselves through fear of the Jews, Nicodemus alone showed himself to them, and said, How can such persons as these enter into the synagogue?

3 The Jews answered him, But how durst thou enter into the synagogue, who wast a confederate with Christ? Let thy lot be along with him in the other world.

4 Nicodemus answered, Amen; so may it be, that I may have my lot with him in his kingdom.

5 In like manner Joseph, when he came to the Jews, said to them, Why are ye angry with me for desiring the body of Jesus of Pilate? Behold, I have put him in my tomb, and wrapped him up in clean linen, and put a stone at the door of the sepulcher:

6 I have acted rightly towards him; but ye have acted unjustly against that just person, in crucifying him, giving him vinegar to drink, crowning him with thorns, tearing his body with whips, and praying down the guilt of his blood upon you.

7 The Jews at the hearing of this were disquieted and troubled; and they seized Joseph, and commanded him to be put in custody before the Sabbath, and kept there till the Sabbath was over.

8 And they said to him, Make confession; for at this time it is not lawful to do thee any harm, till the first day of the week come. But we know that thou wilt not be thought worthy of a burial; but we will give thy flesh to the birds of the air, and the beasts of the earth.

9 Joseph answered, That speech is like the speech of proud Goliath, who reproached the living God in speaking against David. But ye

scribes and doctors know that God saith by the prophet, Vengeance is mine, and I will repay to you evil equal to that which ye have threatened to me.

10 The God whom you have hanged upon the cross, is able to deliver me out of your hands. All your wickedness will return upon you.

11 For the governor, when he washed his hands, said, I am clear from the blood of this just person. But ye answered and cried out, His blood be upon us and our children. According as ye have said, may ye perish for ever.

12 The elders of the Jews hearing these words, were exceedingly enraged; and seizing Joseph, they put him into a chamber where there was no window; they fastened the door, and put a seal upon the lock;

13 And Annas and Caiaphas placed a guard upon it, and took counsel with the priests and Levites, that they should all meet after the Sabbath, and they contrived to what death they should put Joseph.

14 When they had done this, the rulers, Annas and Caiaphas, ordered Joseph to be brought forth.

(In this place there is a portion of the Gospel lost or omitted. which cannot be supplied. It may, nevertheless, be surmised from the occurrence related in the next chapter, that the order of Annas and Caiaphas were rendered unnecessary by Joseph's miraculous escape, and which was announced to an assembly of people.)

CHAPTER X.

1 Joseph's escape. 2 The soldiers relate Christ's resurrection. 18 Christ is seen preaching in Galilee. 21 The Jews repent of their cruelty to him.

WHEN all the assembly heard this (about Joseph's escape), they admired and were astonished, because they found the same seal upon the lock of the chamber, and could not find Joseph.

2 Then Annas and Caiaphas went forth, and while they were all admiring at Joseph's being gone, behold one of the soldiers, who kept the sepulcher of Jesus, spake in the assembly,

3 That while they were guarding the sepulcher of Jesus, there was an earthquake; and we saw an angel of God roll away the stone of the sepulcher and sit upon it;

4 And his countenance was like lightning and his garment like snow; and we became through fear like persons dead.

5 And we heard an angel saying to the women at the sepulcher of Jesus, Do not fear; I know that you seek Jesus who was crucified; he is risen as he foretold;

6 Come and see the place where he was laid; and go presently, and tell his disciples that he is risen from the dead; and he will go before you into Galilee; there ye shall see him as he told you.

7 Then the Jews called together all the soldiers who kept the sepulcher of Jesus, and said to them, Who are those women, to whom the angel spoke? Why did ye not seize them.

8 The soldiers answered and said, We know not who the women were; besides we became as dead persons through fear, and how could we seize those women?

9 The Jews said to them, As the Lord liveth, we do not believe you;

10 The soldiers answering said to the Jews, when ye saw and heard

Jesus working so many miracles, and did not believe him, how should ye believe us? Ye well said, As the Lord liveth, for the Lord truly does live.

11 We have heard that ye shut up Joseph, who buried the body of Jesus, in a chamber, under a lock which was sealed; and when ye opened it, found him not there.

12 Do ye then produce Joseph whom ye put under guard in the chamber, and we will produce Jesus whom we guarded in the sepulcher.

13 The Jews answered and said, We will produce Joseph, do ye produce Jesus. But Joseph is in his own city of Arimathaea.

14 The soldiers replied, If Joseph be in Arimathaea, and Jesus in Galilee, we heard the angel inform the women.

15 The Jews hearing this, were afraid, and said among themselves, If by any means these things should become public, then everybody will believe in Jesus.

16 Then they gathered a large sum of money, and gave it to the soldiers, saying, Do ye tell the people that the disciples of Jesus came in the night when ye were asleep, and stole away the body of Jesus; and if Pilate the governor should hear of this, we will satisfy him and secure you.

17 The soldiers accordingly took the money, and said as they were instructed by the Jews; and their report was spread abroad among all the people.

18 But a certain priest Phinees, Ada a schoolmaster, and a Levite, named Ageus, they three came from Galilee to Jerusalem, and told the

chief priests and all who were in the synagogues, saying,

19 We have seen Jesus, whom ye crucified, talking with his eleven disciples, and sitting in the midst of them in Mount Olivet, and saying to them,

20 Go forth into the whole world, preach the Gospel to all nations, baptizing them in the name of the Father, and the Son, and the Holy Ghost; and whosoever shall believe and be baptized, shall be saved.

21 And when he had said these things to his disciples, we saw him ascending up to heaven.

22 When the chief priests and elders, and Levites heard these things, they said to these three men, Give glory to the God of Israel, and make confession to him, whether those things are true, which ye say ye have seen and heard.

23 They answering said, As the Lord of our fathers liveth, the God of Abraham, and the God of Isaac, and the God of Jacob, according as we heard Jesus talking with his disciples, and according as we saw him ascending up to heaven, so we have related the truth to you.

24 And the three men farther answered, and said, adding these words, If we should not own the words which we heard Jesus speak, and that we saw him ascending into heaven, we should be guilty of sin.

25 Then the chief priests immediately rose up, and holding the book of the law in their hands, conjured these men, saying, Ye shall no more hereafter declare those things which ye have spoken concerning Jesus.

26 And they gave them a large sum of money, and sent other persons along with them, who should conduct them to their own country, that they might not by any means make any stay at Jerusalem.

27 Then the Jews did assemble all together, and having expressed the most lamentable concern said, What is this extraordinary thing which is come to pass in Jerusalem?

28 But Annas and Caiaphas comforted them, saying, Why should we believe the soldiers who guarded the sepulcher of Jesus, in telling us, that an angel rolled away the stone from the door of the sepulcher?

29 Perhaps his own disciples told them this, and gave them money that they should say so, and they themselves took away the body of Jesus.

30 Besides, consider this, that there is no credit to be given to foreigners, because they also took a large sum of us, and they have declared to us according to the instructions which we gave them. They must either be faithful to us or to the disciples of Jesus.

CHAPTER XI.

1 Nicodemus counsels the Jews. 6 Joseph found. 11 Invited by the Jews to return. 19 Relates the manner of his miraculous escape.

THEN Nicodemus arose, and said, Ye say right, O sons of Israel; ye have heard what those three men have sworn by the Law of God, who said, We have seen Jesus speaking with his disciples upon mount Olivet, and we saw him ascending up to heaven.

2 And the scripture teacheth us that the blessed prophet Elijah was taken up to heaven, and Elisha being asked by the sons of the prophets, Where is our father Elijah? He said to them, that he is taken up to heaven.

3 And the sons of the prophets said to him, Perhaps the spirit hath carried him into one of the mountains of Israel, there perhaps we shall find him. And they besought Elisha, and he walked about with them

three days, and they could not find him.

4 And now hear me, O sons of Israel, and let us send men into the mountains of Israel, lest perhaps the spirit hath carried away Jesus, and there perhaps we shall find him, and be satisfied.

5 And the counsel of Nicodemus pleased all the people; and they sent forth men who sought for Jesus, but could not find him; and they returning, said, We went all about, but could not find Jesus, but we have found Joseph in his city of Arimathaea.

6 The rulers hearing this, and all the people, were glad, and praised the God of Israel, because Joseph was found, whom they had shut up in a chamber, and could not find.

7 And when they had formed a large assembly, the chief priests said, By what means shall we bring Joseph to us to speak with him?

8 And taking a piece of paper, they wrote to him, and said, Peace be with thee, and all thy family, We know that we have offended against God and thee. Be pleased to give a visit to us, your fathers, for we were perfectly surprised at your escape from prison.

9 We know that it was malicious counsel which we took against thee, and that the Lord took care of thee, and the Lord himself delivered thee from our designs. Peace be unto thee, Joseph, who art honorable among all the people.

10 And they chose seven of Joseph's friends, and said to them, When ye come to Joseph, salute him in peace, and give him this letter.

11 Accordingly, when the men came to Joseph, they did salute him in peace, and gave him the letter.

12 And when Joseph had read it, he said, Blessed be the Lord God, who didst deliver me from the Israelites, that they could not shed my blood. Blessed be God, who hast protected me under thy wings.

13 And Joseph kissed them, and took them into his house. And on the morrow, Joseph mounted his ass, and went along with them to Jerusalem.

14 And when all the Jews heard these things, they went out to meet him, and cried out, saying, Peace attend thy coming hither, father Joseph.

15 To which he answered, Prosperity from the Lord attend all the people.

13 And they all kissed him; and Nicodemus took him to his house, having prepared a large entertainment.

17 But on the morrow, being a preparation-day, Annas, and Caiaphas, and Nicodemus, said to Joseph, Make confession to the God of Israel, and answer to us all those questions which we shall ask thee;

18 For we have been very much troubled, that thou didst bury the body of Jesus; and that when we had locked thee in a chamber, we could not find thee; and we have been afraid ever since, till this time of thy appearing among us. Tell us therefore before God, all that came to pass.

19 Then Joseph answering, said Ye did indeed put me under confinement, on the day of preparation, till the morning.

20 But while I was standing at prayer in the middle of the night, the house was surrounded with four angels; and I saw Jesus as the brightness of the sun, and fell down upon the earth for fear.

21 But Jesus laying hold on my hand, lifted me from the ground, and the dew was then sprinkled upon me; but he, wiping my face, kissed me, and said unto me, Fear not, Joseph; look upon me for it is I.

22 Then I looked upon him, and said, Rabboni Elias! He answered me, I am not Elias, but Jesus of Nazareth, whose body thou didst bury.

23 I said to him, show me the tomb in which I laid thee.

24 Then Jesus, taking me by the hand, led me unto the place where I laid him, and showed me the linen clothes, and napkin which I put round his head. Then I knew that it was Jesus, and worshipped him, and said; Blessed be he who cometh in the name of the Lord.

25 Jesus again taking me by the hand, led me to Arimathaea, to my own house, and said to me, Peace be to thee; but go not out of thy house till the fortieth day; but I must go to my disciples.

CHAPTER XII.

1 The Jews astonished and confounded. 16 Simeon's two sons, Charinus and Lenthius, rise from the dead at Christ's crucifixion. 19 Joseph proposes to get them to relate the mysteries of their resurrection. 21 They are sought and found, 22 brought to the synagogue, 23 privately sworn to secrecy, 25 and undertake to write what they had seen.

WHEN the chief priests and Levites heard all these things, they were astonished, and fell down with their faces on the ground as dead men, and crying out to one another, said, What is this extraordinary sign which is come to pass in Jerusalem? We know the father and mother of Jesus.

2 And a certain Levite said, I know many of his relations, religions

persons, who are wont to offer sacrifices and burnt-offerings to the God of Israel, in the temple, with prayers.

3 And when the high-priest Simeon took him up in his arms, he said to him, Lord, now lettest thou thy servant depart in peace, according to thy word; for mine eyes have seen thy salvation, which then halt prepared before the face of all people; a light to enlighten the Gentiles, and the glory of thy people Israel.

4 Simeon in like manner blessed Mary the Mother of Jesus, and said to her, I declare to thee concerning that child; He is appointed for the fall and rising again of many, and for a sign which shall be spoken against;

5 Yea, a sword shall pierce through thine own soul also, and the thoughts of many hearts shall he revealed.

6 Then said all the Jews, Let us send to those three men, who said they saw him talking with his disciples in mount Olivet.

7 After this, they asked them what they had seen; who answered with one accord, In the presence of the God of Israel we affirm, that we plainly saw Jesus talking with his disciples in Mount Olivet, and ascending up to heaven.

8 Then Annas and Caiaphas took them into separate places, and examined them separately; who unanimously confessed the truth, and said, they had seen Jesus.

9 Then Annas and Caiaphas said "Our law saith, By the mouth of two or three witnesses every word shall be established."

10 But what have we said? The blessed Enoch pleased God, and was translated by the word of God; and the burying-place of the blessed Moses is known.

11 But Jesus was delivered to Pilate, whipped, crowned with thorns, spit upon, pierced with a spear, crucified, died upon the cross, and was buried, and his body the honorable Joseph buried in a new sepulcher, and he testifies that he saw him alive.

12 And besides, these men have declared, that they saw him talking with his disciples in Mount Olivet, and ascending up to heaven.

13 Then Joseph rising up, said to Annas and Caiaphas, Ye may be justly under a great surprise, that you have been told, that Jesus is alive, and gone up to heaven.

14 It is indeed a thing really surprising, that he should not only himself arise from the dead, but also raise others from their graves, who have been seen by many in Jerusalem.

15 And now hear me a little We all knew the blessed Simeon, the high-priest, who took Jesus when an infant into his arms in the temple.

16 This same Simeon had two sons of his own, and we were all present at their death and funeral.

17 Go therefore and see their tombs, for these are open, and they are risen: and behold, they are in the city of Arimathaea, spending their time together in offices of devotion.

18 Some, indeed, have heard the sound of their voices in prayer, but they will not discourse with anyone, but they continue as mute as dead men.

19 But come, let us go to them, and behave ourselves towards them with all due respect and caution. And if we can bring them to swear, perhaps they will tell us some of the mysteries of their resurrection.

20 When the Jews heard this they were exceedingly rejoiced.

21 Then Annas and Caiaphas, Nicodemus, Joseph, and Gamaliel, went to Arimathaea, but did not find them in their graves; but walking about the city, they found them on their bended knees at their devotions:

22 Then saluting them with all respect and deference to God, they brought them to the synagogue at Jerusalem; and having shut the gates, they took the book of the law of the Lord,

23 And putting it in their hands, swore them by God Adonai, and the God of Israel, who spake to our fathers by the law and the prophets, saying, If ye believe him who raised you from the dead, to be Jesus, tell us what ye have seen, and how ye were raised from the dead.

24 Charinus and Lenthius, the two sons of Simeon, trembled when they heard these things, and were disturbed, and groaned; and at the same time looking up to heaven, they made the sign of the cross with their fingers on their tongues,

25 And immediately they spake, and said, Give each of us some paper, and we will write down for you all those things which we have seen. And they each sat down and wrote, saying:--

CHAPTER XIII.

1 The narrative of Charinus and Lenthius commences. 3 A great light in hell. 7 Simeon arrives, and announces the coming of Christ.

O LORD Jesus and Father, who art God, also the resurrection and life of the dead, give us leave to declare thy mysteries, which we saw after death, belonging to thy cross; for we are sworn by thy name.

2 For thou hast forbidden thy servants to declare the secret things,

which were wrought by thy divine power in hell.

3 When we were Placed with our fathers in the dept of hell, in the blackness of darkness, on a sudden there appeared the color of the sun like gold, and a substantial purple-colored light enlightening the place.

4 Presently upon this, Adam, the father of all mankind, with all the patriarchs and prophets, rejoiced and said, That light is the author of everlasting light, who hath promised to translate us to everlasting light.

5 Then Isaiah the prophet cried out and said, This is the light of the Father, and the Son of God, according to my prophecy, when I was alive upon earth.

6 The land of Zabulon, and the land of Nephthalim, beyond Jordan, a people who walked in darkness, saw a great light; and to them who dwelled in the region of the shadow of death, light is arisen. And now he is come, and hath enlightened us who sat in death.

7 And while we were all rejoicing in the light which shone upon us, our father Simeon came among us, and congratulating all the company, said, Glorify the Lord Jesus Christ the Son of God.

8 Whom I took up in my arms when an infant in the temple, and being moved by the Holy Ghost, said to him, and acknowledged, That now mine eyes have seen thy salvation, which thou hast prepared before the face of all people; a light to enlighten the Gentiles, and the glory of thy people Israel.

9 All the saints who were in the depth of hell, hearing this, rejoiced the more.

10 Afterwards there came forth one like a little hermit, and was asked by every one, Who art thou?

11 To which he replied, I am the voice of one crying in the wilderness, John the Baptist, and the prophet of the Most High, who went before his coming to prepare his way, to give the knowledge of salvation to his people for the forgiveness of sins.

12 And I, John, when I saw Jesus coming to me, being moved by the Holy Ghost, I said, Behold the Lamb of God, behold him who takes away the sins of the world.

13 And I baptized him in the river Jordan, and saw the Holy Ghost descending upon him in the form of a dove, and heard a voice from heaven saying, This is my beloved Son, in whom I am well pleased.

14 And now while I was going before him, I came down hither to acquaint you, that the Son of God will next visit us, and, as the day-spring from on high, will come to us, who are in darkness and the shadow of death.

CHAPTER XIV.

1 Adam causes Seth to relate what he heard from Michael the archangel, when he sent him to Paradise to entreat God to anoint his head in his sickness.

BUT when the first man our father Adam heard these things, that Jesus was baptized in Jordan, he called out to his son Seth, and said,

2 Declare to your sons, the patriarchs and prophets, all those things, which thou didst hear from Michael, the archangel, when I sent thee to the gates of Paradise, to entreat God that he would annoint my head when I was sick.

3 Then Seth, coming near to the patriarchs and prophets, said, I, Seth, when I was praying to God at the gates of Paradise, beheld the angel of

the Lord, Michael, appear unto me, saying, I am sent unto thee from the Lord; I am appointed to preside over human bodies.

4 I tell thee, Seth, do not pray to God in tears, and entreat him for the oil of the tree of mercy wherewith to anoint thy father Adam for his head-ache;

5 Because thou canst not by any means obtain it till the last day and times, namely, till five thousand and five hundred years be past.

6 Then will Christ, the most merciful Son of God, come on earth to raise again the human body of Adam, and at the same time to raise the bodies of the dead, and when he cometh he will be baptized in Jordan;

7 Then with the oil of his mercyhe will anoint all those who believe in him; and the oil of his mercy will continue to future generations, for those who shall be born of the water and the Holy Ghost unto eternal life.

8 And when at that time the most merciful Son of God, Christ Jesus, shall come down on earth, he will introduce our father Adam into Paradise, to the tree of mercy.

9 When all the patriarchs and prophets heard all these things from Seth, they rejoiced more.

CHAPTER XV.

1 Quarrel between Satan and the prince of hell, concerning the expected arrival of Christ in hell.

WHILE all the saints were rejoicing, behold Satan, the prince and captain of death, said to the prince of hell,

2 Prepare to receive Jesus of Nazareth himself, who boasted that he was the Son of God, and yet was a man afraid of death, and said, My soul is sorrowful even to death.

3 Besides he did many injuries to me and to many others; for those whom I made blind and lame and those also whom I tormented with several devils, he cured by his word; yea, and those whom I brought dead to thee, he by force takes away from thee.

4 To this the prince of hell replied to Satan, Who is that so powerful prince, and yet a man who is afraid of death?

5 For all the potentates of the earth are subject to my power, whom thou broughtest to subjection by thy power.

6 But if he be so powerful in his human nature, I affirm to thee for truth, that he is almighty in his divine nature, and no man can resist his power:

7 When therefore he said he was afraid of death, he designed to ensnare thee, and unhappy it will be to thee for everlasting ages,

8 Then Satan replying, said to the prince of hell, Why didst thou express a doubt, and wast afraid to receive that Jesus of Nazareth, both thy adversary and mine?

9 As for me, I tempted him and stirred up my old people the Jews with zeal and anger against him;

10 I sharpened the spear for his suffering; I mixed the gall and vinegar, and commanded that he should drink it; I prepared the cross to crucify him, and the nails to pierce through his hands and feet; and now his death is near at hand, I will bring him hither, subject both to thee and me.

11 Then the prince of hell answering, said, Thou saidst to me just now, that he took away the dead from me by force.

12 They who have been kept here till they should live again upon earth, were taken away hence, not by their own power, but by prayers made to God, and their almighty God took them from me.

13 Who then is that Jesus of Nazareth that by his word hath taken away the dead from me without prayer to God?

14 Perhaps it is the same who took away from me Lazarus, after he had been four days dead, and did both stink and was rotten, and of whom I had possession as a dead person, yet he brought him to life again by his power.

15 Satan answering, replied to the prince of hell, It is the very same person, Jesus of Nazareth.

16 Which when the prince of hell heard, he said to him, I adjure thee by the powers which belong to thee and me, that thou bring him not to me.

17 For when I heard of the power of his word, I trembled for fear, and all my impious company were at the same disturbed;

18 And we were not able to detain Lazarus, but he gave himself a shake, and with all the signs of malice he immediately went away from us; and the very earth, in which the dead body of Lazarus was lodged, presently turned him out alive.

19 And I know now that he is Almighty God who could perform such things, who is mighty in his dominion, and mighty in his human nature, who is the Savior of mankind.

20 Bring not therefore this person hither, for he will set at liberty all those whom I hold in prison under unbelief, and bound with the fetters of their sins, and will conduct them to everlasting life.

CHAPTER XVI.

1 Christ's arrival at hell-gates; the confusion thereupon. 19 He descends into hell.

AND while Satan and the Prince of hell were discoursing thus to each other, on a sudden there was a voice as of thunder, and the rushing of winds, saying, Lift up your gates, O ye princes; and be ye lift up, O everlasting gates, and the King of Glory shall come in.

2 When the prince of hell heard this, he said to Satan, Depart from me, and begone out of my habitations; if thou art a powerful warrior, fight with the King of Glory. But what hast thou to do with him?

3 And he cast him forth from his habitations.

4 And the prince said to his impious officers, Shut the brass gates of cruelty, and make them fast with iron bars, and fight courageously, lest we be taken captives.

5 But when all the company of the saints heard this they spake with a loud voice of anger to the prince of hell,

6 Open thy gates, that the King of Glory may come in.

7 And the divine prophet David cried out, saying, Did not I, when on earth, truly prophesy and say, O that men would praise the Lord for his goodness, and for his wonderful works to the children of men!

8 For he hath broken the gates of brass, and cut the bars of iron in

sunder. He hath taken them because of their iniquity, and because of their unrighteousness they are afflicted.

9 After this, another prophet, namely, holy Isaiah, spake in like manner to all the saints, Did not I rightly prophesy to you when I was alive on earth?

10 The dead men shall live, and they shall rise again who are in their graves, and they shall rejoice who are in the earth; for the dew which is from the Lord, shall bring deliverance to them.

11 And I said in another place, O grave, where is thy victory? O death, where is thy sting?

12 When all the saints heard these things spoken by Isaiah, they said to the prince of hell, Open now thy gates, and take away thine iron bars; for thou wilt now be bound, and have no power.

13 Then was there a great voice, as of the sound of thunder, saying, Lift up your gates, O princes; and be ye lifted up, ye gates of hell, and the King of Glory will enter in.

14 The prince of hell perceiving the same voice repeated, cried out, as though he had been ignorant, Who is that King of Glory?

15 David replied to the prince of hell, and said, I understand the words of that voice, because I spake them in his spirit. And now, as I have before said, I say unto thee, the Lord strong and powerful, the Lord mighty in battle: he is the King of Glory, and he is the Lord in heaven and in earth.

16 He hath looked down to hear the groans of the prisoners, and to set loose those that are appointed to death.

17 And now, thou filthy and stinking prince of hell, open thy gates, that the King of Glory may enter in; for he is the Lord of heaven and earth.

18 While David was saying this, the mighty Lord appeared in the form of a man, and enlightened those places which had ever before been in darkness.

19 And broke asunder the fetters which before could not be broken; and with his invincible power visited those who sate in the deep darkness by iniquity, and the shadow of death by sin.

CHAPTER XVII.

1 Death and the devils in great horror at Christ's coming. 13 He tramples on death, seizes the prince of hell, and takes Adam with him to Heaven.

IMPIOUS death and her cruel officers hearing these things, were seized with fear in their several kingdoms, when they saw the clearness of the light,

2 And Christ himself on a sudden appearing in their habitations, they cried out therefore, and said, We are bound by thee; thou seemest to intend our confusion before the Lord.

3 Who art thou, who has no signs of corruption, but that bright appearance which is a full proof of thy greatness, of which yet thou seemest to take no notice?

4 Who art thou, so powerful, and so weak, so great and so little; mean, and yet a soldier of the first rank, who can command in the form of a servant and a common soldier?

5 The king of Glory, dead and alive, though once slain upon the cross?

6 Who layest dead in the grave, and art come down alive to us, and in thy death all the creatures trembled, and all the stars were moved; and now hast thy liberty among the dead, and givest disturbance to our legions?

7 Who art thou, who dost release the captives that were held in chains by original sin, and bringest them into their former liberty?

8 Who art thou, who dost spread so glorious and divine a light over those who were made blind by the darkness of sin?

9 In like manner all the legions of devils were seized with the like horror, and with the most submissive fear cried out, and said,

10 Whence comes it, O thou Jesus Christ, that thou art a man so powerful and glorious in majesty so bright as to have no spot, and so pure as to have no crime? For that lower world of earth, which was ever till now subject to us, and from whence we received tribute, never sent us such a dead man before, never sent such presents as these to the princes of hell.

11 Who therefore art thou, who with such courage enterest among our abodes, and art not only not afraid to threaten us with the greatest punishments, but also endeavourest to rescue all others from the chains in which we hold them?

12 Perhaps thou art that Jesus, of whom Satan just now spoke to our prince, that by the death of the cross thou wert about to receive the power of death.

13 Then the King of Glory trampling upon death, seized the prince of hell, deprived him of all his power, and took our earthly father Adam with him to his glory.

CHAPTER XVIII.

1 Beelzebub, prince of hell, vehemently upbraids Satan for persecuting Christ and bringing him to hell. 14 Christ gives Beelzebub dominion over Satan forever, as a recompense for taking away Adam and his sons.

THEN the prince of hell took Satan, and with great indignation said to him, O thou prince of destruction, author of Beelzebub's defeat and banishment, the scorn of God's angels and loathed by all righteous persons! What inclined thee to act thus?

2 Thou wouldst crucify the King of Glory, and by his destruction, hast made us promises of very large advantages, but as a fool wert ignorant of what thou wast about.

3 For behold now that Jesus of Nazareth, with the brightness of his glorious divinity, puts to flight all the horrid powers of darkness and death;

4 He has broke down our prisons from top to bottom, dismissed all the captives, released all who were bound, and all who were wont formerly to groan under the weight of their torments, have now insulted us, and we are like to be defeated by their prayers.

5 Our impious dominions are subdued, and no part of mankind is now left in our subjection, but on the other hand, they all boldly defy us;

6 Though, before, the dead never durst behave themselves insolently towards us, nor being prisoners, could ever on any occasion be merry.

7 O Satan, thou prince of all the wicked, father of the impious and abandoned, why wouldest thou attempt this exploit, seeing our prisoners were hitherto always without the least hope of salvation and life?

8 But now there is not one of them does ever groan, nor is there the least appearance of a tear in any of their faces.

9 O prince Satan, thou great keeper of the infernal regions, all thy advantages which thou didst acquire by the forbidden tree, and the loss of Paradise, thou hast now lost by the wood of the cross;

10 And thy happiness all then expired, when thou didst crucify Jesus Christ the King of Glory.

11 Thou hast acted against thine own interest and mine, as thou wilt presently perceive by those large torments and infinite punishments which thou art about to suffer.

12 O Satan, prince of all evil, author of death, and source of all pride, thou shouldest first have inquired into the evil crimes of Jesus of Nazareth, and then thou wouldest have found that he was guilty of no fault worthy of death.

13 Why didst thou venture, without either reason or justice, to crucify him, and hast brought down to our regions a person innocent and righteous, and thereby hast lost all the sinners, impious and unrighteous persons in the whole world?

14 While the prince of hell was thus speaking to Satan, the King of Glory said to Beelzebub the prince of hell, Satan the prince shall he subject to thy dominions for ever, in the room of Adam and his righteous sons, who are mine,

CHAPTER XIX.

1 Christ takes Adam by the hand, the rest of the saints join hands, and they all ascend with him to Paradise.

THEN Jesus stretched forth his hand, and said, Come to me, all ye my saints, who were created in my image, who were condemned by the tree of the forbidden fruit, and by the devil and death;

2 Live now by the wood of my cross; the devil, the prince of this world, is overcome, and death is conquered,

3 Then presently all the saints were joined together under the hand of the most high God; and the Lord Jesus laid hold on Adam's hand, and said to him, Peace be to thee, and all thy righteous posterity, which is mine.

4 Then Adam, casting himself at the feet of Jesus, addressed himself to him with tears, in humble language, and a loud voice, saying,

5 "I will extol thee, O Lord, for thou halt lifted me up, and hast not made my foes to rejoice over me. O Lord my God, I cried unto thee, and thou hast healed me."

6 "O Lord thou hast brought up my soul from the grave; thou hast kept me alive, that I should not go down to the pit."

7 "Sing unto the Lord, all ye saints of his, and give thanks at the remembrance of his holiness, for his anger endureth but for a moment; in his favor is life."

8 In like manner all the saints, prostrate at the feet of Jesus, said with one voice, Thou art come, O Redeemer of the world, and hast actually accomplished all things, which thou didst foretell by the law and thy holy prophets.

9 Thou hast redeemed the living by thy cross, and art come down to us, that by the death of the cross thou mightest deliver us from hell, and by thy power from death.

10 O Lord, as thou hast put the ensigns of thy glory in heaven, and hast set up the sign of thy redemption, even thy cross on earth; so, Lord, set the sign of the victory of thy cross in hell, that death may have dominion no longer.

11 Then the Lord stretching forth his hand, made the sign of the cross upon Adam, and upon all his saints.

12 And taking hold of Adam by his right hand, he ascended from hell, and all the saints of God followed him.

13 Then the royal prophet, David, boldly cried, and said, O sing unto the Lord a new song, for he hath done marvelous things; his right hand and his holy arm have gotten him the victory.

14 The Lord hath made known his salvation, his righteousness hath he openly shewn in the sight of the heathen.

15 And the whole multitude of saints answered, saying, This honor have all his saints, Amen, Praise ye the Lord.

16 Afterwards, the prophet Habbakuk cried out, and said, Thou wentest forth for the salvation of thy people, even for salvation with thine anointed.

17 And all the saints said, Blessed is he who cometh in the name of the Lord; for the Lord hath enlightened us. This is our God for ever and ever; he shall reign over us to everlasting ages. Amen.

18 In like manner all the prophets spake the sacred things of his praise, and followed the Lord.

CHAPTER XX.

1 Christ delivers Adam to Michael the archangel. 3 They meet Enoch and Elijah in heaven, 5 and also the blessed thief, who relates how he came to Paradise.

THEN the Lord, holding Adam by the hand, delivered him to Michael the archangel; and he led them into Paradise, filled with mercy and glory;

2 And two very ancient men met them, and were asked by the saints, Who are ye, who have not yet been with us in hell, and have had your bodies placed in Paradise?

3 One of them answering, said, I am Enoch, who was translated by the word of God: and this man who is with me, is Elijah the Tishbite, who was translated in a fiery chariot.

4 Here we have hitherto been, and have not tasted death, but are now about to return at the coming of Antichrist, being armed with divine signs and miracles, to engage with him in battle, and to be slain by him at Jerusalem, and to be taken up alive again into the clouds, after three days and a half.

5 And while the holy Enoch and Elias were relating this, behold there came another man in a miserable figure, carrying the sign of the cross upon his shoulders.

6 And when all the saints saw him, they said to him, Who art thou? For thy countenance is like a thief's; and why dost thou carry a cross upon thy shoulders?

7 To which he answering, said, Ye say right, for I was a thief, who committed all sorts of wicked. ness upon earth.

8 And the Jews crucified me with Jesus; and I observed the surprising

things which happened in the creation at the crucifixion of the Lord Jesus.

9 And I believed him to be the Creator of all things, and the Almighty King; and I prayed to him, saying, Lord remember me, when thou comest into thy kingdom.

10 He presently regarded my supplication, and said to me, Verily I say unto thee, this day thou shalt be with me in Paradise.

11 And he gave me this sign of the cross, saying, Carry this, and go to Paradise; and if the angel who is the guard of Paradise will not admit thee, show him the sign of the cross, and say unto him Jesus Christ who is now crucified, hath sent me hither to thee.

12 When I did this and told the angel who is the guard of Paradise all these things, and he heard them, he presently opened the gates, introduced me, and placed me on the right hand in Paradise,

13 Saying, Stay here a little time, till Adam, the father of all mankind, shall enter in, with all his sons, who are the holy and righteous servants of Jesus Christ, who was crucified.

14 When they heard all this account from the thief, all the patriarchs said with one voice, Blessed be thou, O Almighty God, the Father of everlasting goodness, and the Father of mercies, who hast shown such favor to those who were sinners against him, and hast brought them to the mercy of Paradise, and hast placed them amidst thy large and spiritual provisions, in a spiritual and holy life. Amen.

CHAPTER XXI.

1 Charinus and Lenthius being only allowed three days to remain on earth, 7 deliver in their narratives, which miraculously correspond;

they vanish, 13 and Pilate records these transactions.

THESE are the divine and sacred mysteries which we saw and heard. We, Charinus and Lenthius are not allowed to declare the other mysteries of God, as the archangel Michael ordered us,

2 Saying, ye shall go with my brethren to Jerusalem, and shall continue in prayers, declaring and glorifying the resurrection of Jesus Christ, seeing he hath raised you from the dead at the same time with himself.

3 And ye shall not talk with any man, but sit as dumb persons till the time come when the Lord will allow you to relate the mysteries of his divinity.

4 The archangel Michael farther commanded us to go beyond Jordan, to an excellent and fat country, where there are many who rose from the dead along with us for the proof of the resurrection of Christ.

5 For we have only three days allowed us from the dead, who arose to celebrate the passover of our Lord with our parents, and to bear our testimony for Christ the Lord, and we have been baptized in the holy river of Jordan. And now they are not seen by any one.

6 This is as much as God allowed us to relate to you; give ye therefore praise and honor to him, and repent, and he will have mercy upon you. Peace be to you from the Lord God Jesus Christ, and the Savior of us all. Amen, Amen, Amen.

7 And after they had made an end of writing, and had written on two distinct pieces of paper, Charinus gave what he wrote into the hands of Annas, and Caiaphas, and Gamaliel.

8 Lenthius likewise gave what be wrote into the hands of Nicodemus and Joseph; and immediately they were changed into exceeding white

forms and were seen no more.

9 But what they had written was found perfectly to agree, the one not containing one letter more or less than the other.

10 When all the assembly of the Jews heard all these surprising relations of Charinus and Lenthius, they said to each other, Truly all these things were wrought by God, and blessed be the Lord Jesus for ever and ever, Amen.

11 And they went all out with great concern, and fear, and trembling, and smote upon their breasts and went away every one to his home.

12 But immediately all these things which were related by the Jews in their synagogues concerning Jesus, were presently told by Joseph and Nicodemus to the governor.

13 And Pilate wrote down all these transactions, and placed all these accounts in the public records of his hall.

CHAPTER XXII.

1 Pilate goes to the temple; calls together the rulers, and scribes, and doctors. 2 Commands the gates to be shut; orders the book of the Scriptures; and causes the Jews to relate what they really knew concerning Christ. 14 They declare that they crucified Christ in ignorance, and that they now know him to be the Son of God, according to the testimony of the Scriptures; which, after they put him to death, were examined.

AFTER these things Pilate went to the temple of the Jews, and called together all the rulers and scribes, and doctors of the law, and went with them into a chapel of the temple.

2 And commanding that all the gates should be shut, said to them, I have heard that ye have a certain large book in this temple; I desire you, therefore, that it may be brought before me.

3 And when the great book, carried by four ministers of the temple, and adorned with gold and precious stones, was brought, Pilate said to them all, I adjure you by the God of your Fathers, who made and commanded this temple to be built, that ye conceal not the truth from me.

4 Ye know all the things which are written in that book; tell me therefore now, if ye in the Scriptures have found any thing of that Jesus whom ye crucified, and at what time of the world he, ought to have come: show it me.

5 Then having sworn Annas and Caiaphas, they commanded all the rest who were with them to go out of the chapel.

6 And they shut the gates of the temple and of the chapel, and said to Pilate, Thou hast made us to swear, O judge, by the building of this temple, to declare to thee that which is true and right.

7 After we had crucified Jesus, not knowing that he was the Son of God, but supposing he wrought his miracles by some magical arts, we summoned a large assembly in this temple.

8 And when we were deliberating among one another about the miracles which Jesus had wrought, we found many witnesses of our own country, who declared that they had seen him alive after his death, and that they heard him discoursing with his disciples, and saw him ascending into the height of the heavens, and entering into them;

9 And we saw two witnesses, whose bodies Jesus raised from the dead, who told us of many strange things which Jesus did among the dead, of

which we have a written account in our hands.

10 And it is our custom annually to open this holy book before an assembly, and to search there for the counsel of God.

11 And we found in the first of the seventy books, where Michael the archangel is speaking to the third son of Adam the first man, an account that after five thousand five hundred years, Christ the most beloved son of God was to come on earth,

12 And we further considered, that perhaps he was the very God of Israel who spoke to Moses, Thou shalt make the ark of the testimony; two cubits and a half shall be the length thereof, and a cubit and a half the breadth thereof, and a cubit and a half the height thereof.

13 By these five cubits and a half for the building of the ark of the Old Testament, we perceived and knew that in five thousand years and half (one thousand) years, Jesus Christ was to come in the ark or tabernacle of a body;

14 And so our Scriptures testify that he is the Son of God, and the Lord and King of Israel.

15 And because after his suffering, our chief priests were surprised at the signs which were wrought by his means, we opened that book to search all the generations down to the generation of Joseph and Mary the mother of Jesus, supposing him to be of the seed of David;

16 And we found the account of the creation, and at what time he made the heaven and the earth, and the first man Adam, and that from thence to the flood, were two thousand seven hundred and forty- eight years.

17 And from the flood to Abraham, nine hundred and twelve. And

from Abraham to Moses, four hundred and thirty. And from Moses to David the King, five hundred and ten.

18 And from David to the Babylonish captivity five hundred years. And from the Babylonish captivity to the incarnation of Christ, four hundred years.

19 The sum of all which amounts to five thousand and half (a thousand.)

20 And so it appears, that Jesus whom we crucified, is Jesus Christ the Son of God, and true Almighty God. Amen.

(In the name of the Holy Trinity, thus end the acts of our Savior Jesus Christ, which the Emperor Theodosius the Great found at Jerusalem, in the hall of Pontius Pilate, among the public records; the things were acted in the nineteenth year of Tiberius Caesar, Emperor of the Romans, and in the seventeenth year of the government of Herod, the son of Herod and of Galilee, on the eighth of the calends of April, which is the twenty-third day of the month of March, in the CCIId Olympiad, when Joseph and Caiaphas were rulers of the Jews; being a History written in Hebrew by Nicodemus, of what happened after our Savior's crucifixion.)

You should also know the following apocryphal books...

Both with references inside the Bible...

The Book of

JASHER

Visit Amazon.com and click to LOOK INSIDE

Only US$ 8.95